# RESEARCHES ON LIVING PTERIDOPHYTES IN INDIA, BURMA AND CEYLON

# Researches on Living Pteridophytes in India, Burma and Ceylon

N. P. CHOWDHURY

ASIA PUBLISHING HOUSE
NEW YORK

Nira Pad **Chowdhury** (1911)



ISBN O 210 22849 9

PRINTED IN INDIA
AT THE DELHI UNIVERSITY PRESS, AND PUBLISHED BY P. S. JAYASINGHE
ASIA PUBLISHING HOUSE INC., 118 EAST 59TH STREET, NEW YORK, NY 10022.

*To the loving memory of my brother*
*the late Professor H. P. Chowdhury*

# PREFACE

This brief, comprehensive volume on pteridological studies in India, Burma and Ceylon has been compiled for postgraduate and research-level undertakings. It highlights the contributions on the subject during a century and a quarter ending 1968. The fast growing literature together with the often-encountered difficulty of their ready and timely accessibility has, no doubt, hampered the pooling of essential data for a systematic and chronological integration of the topic into a coherent theme; which establishes the need to make this work a continuing one by bringing out periodical supplements to it. Emphasis has been laid, however, on the preparation of an exhaustive bibliography. I hope this treatise will serve as a baseline for a further growth of work in this direction.

In writing this book, I have drawn freely from a review of the pteridological researches in India, Burma and Ceylon which I had to write for the Indian Botanical Society in their erstwhile History of Botanical Researches series. I have utilised also, to a considerable extent, my earlier contribution (1957) on 'Pteridology' published by the National Institute of Sciences of India.

I wish to record here my grateful thanks to Professor P. N. Mehra of Punjab University, Professor T. S. Mahabale of Poona University, Professor A. Abraham of Kerala University and the late Reverend H. Santapau of the Botanical Survey of India for their very helpful and encouraging attitude towards this work. I am indebted to Mrs. S. Sahni of Birbal Sahni Palaeobotanical Institute, Lucknow and Professor B. M. Johri of Delhi University for the facilities given to consult the reprint libraries of the late Professor Birbal Sahni, Sc. D., F. R. S. and the late Professor P. Maheswari, D. Sc., F. R. S. of Lucknow and Delhi respectively. I am also indebted to other colleagues, notably Dr. Vishnu Mitre (Palaeobotanical Institute, Lucknow), Dr. M. L. Bannerji (Kalyani University), Dr. S. K. Roy (Benares Hindu University), Dr. S. Bhambi (Birla Institute of Technology & Science, Pilani), Miss Usha Sharma

(Lucknow University), the late Professor G. P. Majumdar (Calcutta University), Dr. P. Kachroo (Gauhati University), Dr. B. K. Nayar (National Botanic Gardens, Lucknow), Professor K. M. Gupta (Jodhpur University), Dr. R. N. Kapil (Delhi University), Dr. J. P. Correa and Dr. H. L. Chakravarty (both of UNESCO). And I must acknowledge the unfailing assistance I have received from my son, S. P. Chowdhury, in the correction of proofs and in compiling the index.

*November 1970,*
*Department of Botany,*
*University of Delhi.*

N. P. CHOWDHURY

# INTRODUCTION

Earlier works on Indian Pteridophytes, as on other groups of plants, have been mostly in the nature of systematic, geographic and economic studies done chiefly by the Botanical Survey of India (i.e. Botanical Departments of Bengal, Madras, Bombay and Northern India) and the Forest Services of India. Thus, Royle's (1832) [1] enumeration of Filices, Marsiliaceae, Equisetaceae and Lycopodinae of the Saharanpur herbarium; Beddome's (1868, 1883-92) Ferns of India ; Hope's (1890) *Lastreas* from Assam, Hope and Trotter's (1890) Catalogue of Ferns at Saharanpur Herbarium ; and Blanford's (1888)[2] A List of the Ferns of Simla are some of the earliest contributions. Spring's monograph (1841, 1848) on Lycopodiaceae and Baker's (1887) Handbook of Fern Allies include taxonomic accounts of some of the lower pteridophytes of India, Burma and Ceylon. In the economic sphere, Sir George Watt's (1896, 1908) Dictionary of Economic Products of India gives an account of some economic species like the Male Fern, *Lycopodium*, and Kirtikar and Basu's (1941, Vol. 4) Indian Medicinal Plants mentioned medicinal utility of some of the Polypodiaceae, Ophioglossaceae, and *Equisetum debile*.

In the first decade of the present century, floristic investigations in different parts of India continued to form the major botanical studies, and Pteridophyta taxonomy, naturally, found place in some of these, notable among which are Bengal Plants by D. Prain (1903), Botany of Bihar and Orissa by H. H. Haines (1924), and J. L. Stewart's (1869) Punjab Plants (Lahore). Next in bulk, after the floristic or taxonomic contributions on Pteridophytes, come the general morphological and anatomical contributions on their sporophytes such as are seen in Sahni's (1924) monograph on *Tmesipteris*, Chowdhury's (1933, 1937) on *Lycopodium* or Kashyap's (1917) and Pathak's (1935) on *Equisetum*, Vashist's (1927) *Ophioglossum* species ; Hansraj's (1923) *Lygodium* ; Ekambaram's (1933) *Iseotes* and P. Maheswari and B. Singh's (1934) *Ophoiglossum fibrosum* In the life history work on Pteridophytes of India, Burma and Ceylon very little has been done worth mentioning, almost the only contributions being those of Sethi (1928) on *Equisetum debile*, Sampat Kumaran and Gopala Iyengar (1932) on *Gleichenia dichotoma*, Rao (1935) on *Azolla* of Lucknow, and Kolhatkar (1937) on *Marsilia* of Poona.

Very fragmentary contributions are met with on the gametophytes of Indian Pteridophyta. In India, however, *Equisetum* and some higher ferns have received most attention in this direction than others of the class. In the lower Pteridophytes, Sethi (1926), Kashyap (1932), Pathak (1931, 1932), and, recently, Chatterji and Ram (1968) described *Equisetum* prothalii. Mahabale (1937, 1948), for the first time in India, described the prothalli of *Lycopodium cernuum* and *L. hamiltonii* and Ekambaram and Venkatanathan (1933) that of *Iseotes coromandelina*. Among the ferns, these have been studied in *Ophioglossum* species of Poona by Mahabale (1937), *Azolla pinnata* by Mullay (1928), and Rao (1935) in *Diplazium esculentum* (1948), and *Dryopteris* (1931, 1948) by Mehra, *Actiniapteris australis* by Stokey (1948), *Goniopteris* by Mahabale (1947) and Mehra (1952), also those of *Gymnopteris variabilis* (1947) and *Ceratopteris thallictroides* (1948) by Mahabale.

Some of the earlier cytological investigations of importance done in India are by Mehra (1937, 1945) on *Adiantum lunulatum;* by Sarbadhikari (1924, 1927, 1935) on *Osmunda* and *Doodia;* by Mulay (1937) on chromatin of *Azolla pinnata*. More recent investigations are by Abraham and Ninan (1954) on *Ophioglossum reticulatum;* Ninan (1955, 1956 and 1958) on *Equisetum*, *Psilotum*, *Lycopodium* and *Isoetes;* Mahabale and Chenna (1954) on *Regnelledium* and *Marsilia;* Roy (1964-65) on *Mohria*, *Dicksonia;* and Manton and Sledge (1954) on Pteridophyte Flora of Ceylon.

Finally, the studies from the point of view of an assessment of Phylogenetical and evolutionary importance of characters have been extremely stray and meagre, the pioneer work in this being that of Professor Birbal Sahni on branching of filicales (1917), *Psilotum* (1923), *Tmesipteris* (1921), also ontogeny of vascular plants (1925), and Sinha (1923) on the evolution of archegonium. Of late, a few contributions came from Chowdhury (1932) on evolutionary sporangial feature of

Osmundaceae, Mahabale (1945, 1963) on concept of vestigial organs and evolutionary tendencies in *Azolla* and Dickason (1946) in a phylogenetic study of the ferns of Burma.

The bulk of the contemporary contributions on Pteridophyta of India, Burma, and Ceylon, however, again seems to be of the floristic nature, namely, Mehra (1929), Ferns of Mussoorie ; Raizada and Chowdhury (1954, 1961), Ferns and Fern Allies of the Upper Gangetic Plain; Stewart (1938, 1945, 1951, 1952 and 1957), Ferns of Dehra Dun, Mussoorie, Pahalgam (Kashmir), and Rawalpindi in West Pakistan ; Kachroo (1953, 1961), Assam Ferns ; Mahabale (1931), Blatter, and De Almeida's (1922) accounts of Bombay Presidency Pteridophytic Vegetation; Mehra and Bir (1964), Pteridophytic Flora of Darjeeling and Sikkim (Himalayas), and Panigrahi and co-workers (1962, 1966, and 1968) are some recent contributions in the line. Dickason (1946) studied ferns of Burma; Trimen (1885, 1900) and Abeywickrama (1956) contributed on Pteridophytes of Ceylon.

Not all groups of Pteridophyta seem to have been adequately studied in this part of the globe and the families like Cyatheaceae, Parkeriaceae, Gleicheniaceae, Marattiaceae, Hymenophyllaceae and some less known genera of Polpodiaceae, no doubt, demand more attention of the pteridologists here. Also the new avenues of research like tissue culture, cytotaxonomy, genetics (hybridization) of Bucholtz and Sellett (1941)[3] and morphogenesis of pteridophyta and the culture of their gametophytes under varying experimental conditions appear to offer vast fields of unexplored deliberations of morphological or even evolutionary interest. The economic utility of pteridophytes, their biochemistry for an assessment of drug value and alkaloid content, if any, are other avenues which demand most careful consideration.

[1] Royle, J. F. (1832) : An account of the Honourable Company's Botanic Gardens at Saharanpur. Journ. Asiatic Soc. Bengal I (2), 41-58.

[2] Blanford, H. F. (1888) : A List of the Ferns of Simla in N. W. Himalayas. Journ. Asiatic Soc. Bengal LVII (4).

[3] Bucholtz, J. T. and Selett, J. W. (1941) : The hybridization of water ferns—*Marsilia* and *Pilulania*, Naturalist, Vol. 75, pp. 90-97, 1941. *(M. quudrifolia* X *M. crenata.)*

# CONTENTS

# LIST OF PLATES

# I PSILOTACEAE

The majority of Indian contribution on *Psilotum* pertains to the mere recording of the distribution of *P. nudum* (L.) Beauv. (Syn. *P. triquetrum* Sw.) in the various regions of the country, thus Joshi (1935), Mahabale (1938) and Raizada and Chowdhury (1954, 1961) mentioned its occurrence in the provinces of the Punjab (Kulu), Bombay, and U. P. respectively. Similarly, Venkateshwarulu (1944) and B. S. M. Dutt (1950) recorded it from the Godavari District and South India. Raizada (1935) gave its distribution in India. In its morphological studies, Sahni (1923) discussed the significance of certain sporangiophore abnormalities and the archaic features of *Psilotum* while T.C.N. Singh (1930) described medullary tracheids in *P. triquetrum*, and compared the condition with that of *Tmesipteris vieillardi*.

Sahni (1921) discussed a sporophyll abnormality in *Tm. vieillardi*, i.e. trilocular synangia both on two-lobed and three-lobed sporophylls and also gave a structural and, later on, a monographic morphological account (1925) of this New Caledonian species. Sahni (1925) observed that *Tmesipteris vieillardi* Dangeard in its erect habit, terrestrial mode of life and presence of medullary xylem represents the most primitive or the archaic form of the genus, perhaps indicating the parent type which gave rise to the usual pendulous epiphytic forms grouped under *Tm. tannensis* Bernh. The medullary xylem in the aerial shoot of *Tm. vieillardi* is most comparable with the central cauline xylem in the stele of *Asteroxylon* and *Lycopodium*. Sahni noted that in the aerial shoot of *Tmesipteris* the phloem is not confined to a layer outside the xylem ring, as is commonly described, but surrounded each individual bundle of the ring; also the sporangiophores of the Psilotales are not forked sporophylls but are axial organs comparable with the branched sporangium tipped axes of the Psilophytales as also suggested by Kidston and Lang (1917-21). An indication of verticillate organization is revealed by the mode of branching of sporangiophores and alternate occurrence of the sterile lobes comparable with Sphenophyllales and Equisetales.

Sahni (1925) regarded the presence of cauline medullary xylem in the basal or first formed part of the axis in *Tm. vieillardi* as of the nature of recapitulation character, just as Holloway (1917 ex Sahni 1925) had found that the seedlings of *Tmesipteris* possessed a protostele instead of the ring of bundles seen in the mature stem.

C. A. Ninan (1956) investigated the cytology of *Psilotum nudum* (L.) Beauv. from India and Ceylon and found only tetraploids and for the first time a triploid. No diploid is yet reported from India. With n = 52, tetravalents (in tetraploids) and trivalents (in triploids) were seen during meioses of the spore mother cells and are indicative of natural autopolyploidy in the genus, and probable extinction of the diploid race. A base number of thirteen chromosomes is suggested for the primitive pteridophytes giving rise to multiples like 52, 104, 156 and 208. Pant and Mehra (1963) found haplocheilic development of stomata in *Psilotum nudum.*

No gametophytic study has been made in India of Psilotales nor is there any record so far of the occurrence of *P. complanatum* (Syn. *P. flaccidum*) or *Tmesipteris* from India. Dalgado (1896) and Baker (1887) were one of the earliest to mention the existence of *Psilotum* in India.

Abeywickrama (1956) reported the occurrence of *P. nudum* (L.) Beauv. in wet zone areas of Ceylon, from sea level to highest elevation, as a rare plant only.

## II LYCOPODIACEAE

On this group, very little work has been done in India. In some of the early foreign works by Spring (1841, 1848), Baker (1887), and Pritzell (1902), morphological and systematic studies of some species—which occur in India—are found. The only exclusive contribution is by N. P. Chowdhury (1933, 1937) who gave the distribution of about 30 species (1937) and discussed the anatomy, modes of vegetative reproduction and the epidermal structure of some of them with ecological and taxonomic observations. Chowdhury (1937) also described certain unusual stelar variations of the cortical roots which, besides the usual crescentic xylem form, showed also annular (*L. hamiltonii*), parallel banded or irregular shaped (*L. phlegmaria*) structures. Peculiar-amphivasal stem stele of *L. hamiltonii* and the anatomy of the roots tubercle of *L. cernuum* were also described by him. Mahabale (1937, 1948) described the prothalli of *L. cernuum* from Karanzol and *L. hamiltonii* from North Kanara. The prothallus of the latter has a central conical portion bearing archegonia and several lateral branches with antheridia in contrast to the cauliflower head-shaped ; green, leafy lobes crowned prothallus of *L. cernuum*, which had a broad rounded base bearing rhizoids, with the antheridia and archegonia occurring below the green lobes. Ghosh (1942) recorded double sporangiate sporophylls in *L. phlegmaria*, while Banerji and Majumdar (1952) studied the origin and development of the shoot apex in the same species.

Ninan (1958) studied the cytology of nine species of *Lycopodium* of S. India. The species possess chromosome numbers : n=48 (*L. wightianum* Wall.), n=104 (*L. cernuum* L.), n=136 (*L. hamiltonii* Spr., *L. vernicosum* Hook. et Grev., *L. squarrosum* Forst, *L. setaceum* Hamilt , *L. macrostachys* L., *L. phlegmaria* L.), and 2n=c 405 (*L. lucidulum* Michx.). A single genus having more than one basic number, such as 11, 12, 13, and 17 in its different species, is remarkably peculiar to *Lycopodium*, and is not found in other genera of primitive pterodophytes. Vegetatively propagative species like *L. selago*, *L. lucidulum* (probably a hybrid) show irregular meiotic behaviour comparable to *Phylloglossum drumondii*.

Bhambi (1965) studied the development, structure, and arrangement of leaves in *L. complanatum*, *L. clavatum*, *L. obscurum*, *L. nikoense*, *L. squarrosum*. The leaf originating from a group of superficial cells has a mesarch vascular bundle and haplocheilic stomata. Leaf size and arrangement vary from 2mm —12mm to 30 mm in length with phyllotaxy 2/5—2/13, opposite (*L. complanatum*) or whorled (*L. nikoense*). Leaf dimorphism is seen in *L. obscurum*, *L. complanatum*. Bhambi and Puri (1963) observed no definite apical cell in the shoot apex of the *Lycopodium*. Chowdhury (1937) reported a group of apical cells in the stem apex of (*L. hamiltonii, L. setaceum, L. serratum*).

Mehra and Varma (1957) found the chromosome number in *Lycopodium lucidulum*, n = 132, and *L. clavatum*, n = 34.

Sahni (1925) in his presidential address on recapitulation in vascular plants considered heterophylly as derived from homophyllous condition citing Holloway (1915 ex Sahni, 1925) who showed that seedlings of some heterophyllous species of *Lycopodium* bore leaves of one kind only, the heterophyllous conditions in them arose subsequently; also the radial type of stele of *Lycopodium* is considered more primitive than its parallel banded stele as this type is seen in the seedlings of all species including those of *clavata* and *cernua* sections. Sahni (1925) further thought that Treub's (1890) view of the protocorm of *Lycopodium cernuum*, being a vestigial organ of great antiquity, the precursor of the leafy shoot in ancestral pteridophytes, has materialized in the discovery of Psilophytales in which a closely comparable organ persisted to the adult stage, as also it does in the modern Psilotaceae (Kidston and Lang, Trans. R. S. Edinb. 52 : 619, 846).

Banerji (1962) reported the occurrence of the fungus *Venturia* cladospium stage on *Lycopodium annotinum* L. collected from E. Nepal. Branched intracellular brownish hyphae infecting stem cortex, leaf, sporophyll, and sporangia, produced one to two-celled conidia.

Mehra and Bir (1964) enumerated 16 species of *Lycopodium* from the Darjeeling-Sikkim region of which 7 species, e.g. *L. lucidulum, L. serratum, L. setaceum*, (epiphytic), *L. squarrosum* (epiphytic) *L. phlegmaria* (rare, epiyhytic), *L. cernuum* and *L. clavatum* were collected by the authors.

Abeywickrama (1956) found four species in Ceylon, e.g., *L. pulcherrimum* Wall. ex Hook. et Grev., *L. phlegmaria* L. (both epiphytes), *L. carolianum* (dimorphic leaves) in boggy places of Nuwara Eliya, and *L. cernuum* L., with trailing stems and erect branches are very common from sea level to the highest elevation.

Chowdhury (1937) recorded eight species of *Lycopodium* from Burma and sixteen species from Ceylon, viz *L. hamiltonii* Spr., (Moulmein), *L. aloifolium* (Parpo, Pegu), (*L. serratum*

Thunb. (Nab Tung), *L. squarrossum* Forst (Mergui, Shan States), *L. phlegmaria* L. (Mergui, Tenasserim), *L. cernuum* L. (Upper Burma), *L. clavatum* L. (Moulmein, Malin), *L. casuarinoides* Spr. (Upper Burma); and *L. selago* L., *L. hamiltonii* Spr., *L. aloifolium*, *L. vernicosum*, *L. ceylanicum* Spr., *L. serratum* Thunb., *L. setaceum* Hamilt. (Newera Ellia), *L. subulifolium* Wall., *L. squarrosum* Forst, *L. phlegmaria* L., *L. phyllanthum* Bak., *L. cernuum* L., *L. clavatum* L., *L. carolinianum* L. (Pedro Tallugalla), *L. wightianum* Wall., *L. fawcettii* Underw. (from Ceylon mountains), respectively. He (1937) also recorded thirty-three species of *Lycopodium* from India.

## III SELAGINELLACEAE

Bancroft (1914) described the vegetative reproduction by tubers in two Indian species of *Selaginella* (*S. chrysocaulos*, and *S. chrysorhizos*). S. L. Ghosh (1916, 1932) observed proliferation in the cone of a *Silaginella* species from Garhwal and he (1920) discussed the evolution in the cone of the genus, the radial and dorsiventral types as exemplified by *S. spinosa* and *S. pallidissima.*

Bhambi and Puri (1963) observed a group of apical cells, but no indication of a three or four-sided single apical cell in the shoot apex of *Selaginella* species. Chowdhury (1959) observed a wedge-shaped, two-sided apical cell in the cone apex of *S. bryopteris* (L.) Baker.

G. P. Majumdar described (1931, 1942) the origin of medullation or siphonostele in three species of *Selaginella*: *S. inequalifolia*, *S. wallichii*, *S. canaliculata*. Chowdhury (1943, 1959) studied the anatomy of two xerophytic Indian species, *S. bryopteris* from Banda (U.P.) and *S. Wightii* (1943, 1957) from Nandi Hills, S. India. The monostelic resurgent curly stemmed *S. bryopteris* showed a horse-shoe-shaped or dorsiventrally flattened, ribbon-shaped protostele with two or more marginal protoxylem groups, a broad cortex of thick-walled turgid cells on the dorsal and thin parenchyma, with a narrow sclerenchyma flange on ventral face, leaves (heterophyllous) with thick warty cuticle, and a double layer of well-developed pallisade mesophyll. *S. wightii* had essentially similar monostelic feature but with extensive all-round sclerenchyma development in the cortex and homophyllous leaves with a much thicker warty cuticle deposit and a homogeneous reticulate mesophyll. The species were devoid of rhizophores in the matured stage and exhibited extreme xerophytism. The author also found *S. bryopteris* being sold by local herbalists under the name of 'Sanjivani,' 'Hanspadi,' etc., having considerable medicinal importance according to the Ayurvedic system. Alston (1945) enumerated 57 species of *Selaginella* from India and regions now in Pakistan. P. Maheswari and V. Puri (1935) noted a smut on Indian *Selaginella*. A. K. Mitra (1944) reported a new parasite of *Selaginella*, e g. *Melanopsamma ranjanii*. T.C.N. Singh (1930) found a smut *Entyloma polysporium* on *S. chrysocaulos*. Raizada and Chowdhury (1954, 1961) recorded four species, e.g. *S. bryopteris* Bak., *S. adunca* A. Br., *S. chrysocaulos* Hk. and Grev., and *S. subdiaphana* (Wall.) Spr. from the upper Gangetic plain.

Panigrahi and Dixit (1966) worked out the systematics of *Selaginella* species of India.

Mehra and Bir (1964) recorded nine species of *Selaginella* from Darjeeling and N. Sikkim. *S. involvens*, *S. chrisorhizos*, *S. chrysocaulos*, and *S. monospora* were quite common.

Alston (1945) recorded ten species of *Selaginella* from Ceylon. Abeywickrama (1956) reported *S. wightii* Hieron. from Dambulla (cf. Chowdhury, 1957), *S. omithopodiodes* (L.) Spr. (endemic to Ceylon in the moist zones), *S. crassipes* Spr. (upper montane region) in Ceylon.

## IV ISOETACEAE

Venkatanathan (1933) studied the anatomy of *Iseotes*[1] *coromandelina* L. f. The primary tissues of the stem are differentiated from the apex consisting of a group of uniform cells. Both primary and secondary phloem are found, the cambium cut off "prismatic layer" consisting of secondary phloem and parenchyma only. Like *I. japonica* no secondary traeheids are formed.

[1]*Isoetes coromandelina* L. f., the species first recorded from the Coromandel coast of India, reportedly also occurs in Ceylon (Dambulla-collector Trimen, 1882, ex Abeywickrama, 1956).

Ekambaram and Venkatanathan (1933) studied the sporangial structure and its development in *Iseotes coromandelina*, and found the haploid chromosome number to be 16. Misra (1935) studied the occurrence of *Iseotes* in the Banares region, and Bharadwaj (1935) in the rest of India. Mahabale (1938) recorded a new species, *I. sahyadrii* found at Sahyadris (Mysore State) and Panchgani (Bombay). L. Narayana Rao (1944) described new species of *Iseotes* from Bangalore, while Shende and Mahabale (1947) founded and described another new species *I. dixitii* Shende from Panchgani. Raizada and Chowdhury (1954, 1961) reported *I. coromandelina* from the upper Gangetic Plain in Mirzapur District, U.P. Venkatanathan (1931, 1932) studied megaspore mother cells and microsporangia of *I. coromandelina.* Gaekwad (1954) reported bisporangiate (i.e. containing both micro and mega spores) sporophylls in a species from Baroda. The present author also came across a similar condition in a species of *Iseotes* from S. India. Usha Sharma (1958, 1959) described the vegetative anatomy and megasporogenesis in *I. sampatkumarani.* Only phloem elements were seen to occur in the prismatic tissue and megasporangium, arising from a row of superficial cells, produced dimorphic megaspores.

Bhambi (1957), Bhambi and Puri (1963) found a group of apical meristematic cells in the shoot apex of *Isoetes* (*I. coromandelina*, *I. engelmanni*, *I. japonica*). Usha Sharma (1961) observed similar features in *I. sampatkumarani*. Bhambi (1962, 1963) studied the axis anatomy and structure of root, leaf, and sporophylls of *I. coromandelina* (an apogamous species) together with some other species. Stem is protostelic with "prismatic tissue" of secondary phloem and parenchyma or included pitted tracheids also. Root promeristem showed two regions, a central core forming the stele over-topped by a meristematic mantle forming the cortex, epidermis and root cap. The root, traversed by a monarch exarch stele, branched dichotomously. Leaf originated from a group of superficial cells (ligule from a single cell) and possess haplocheilic stomata.

## V EQUISETACEAE

J. Milde (1867) in his Monographia Equisetorium was one of the earliest to record several Indian species and the present author had the privilege to see Milde's notations on the herbarium sheets of Indian *Equisetum* preserved at the Forest Research Institute, Dehra Dun. Kashyap (1914, 1917, 1930, 1932) studied the sporophytic and gametophytic structures of *E. debile* Roxb., some of his observations being the unstability or variations of endodermis layer (1917); peculiar abnormality of superimposed cones bearing 2-3 whorls of annulus and the topmost cone consisting of part leaf and part sporangiophore in *E. debile* (1930) collected from the Ravi banks (Lahore)., also its protogynous gametophyte with radial symmetry or only male prothalli from spores when lying crowded (1917).

Sethi (1926, 1928) studied the development of archegonium, antheridium, fertilization, embryo development, and general life history of *E. debile*. Phatak (1931, 1932, 1935) studied gametophytic structure, meristem, sex differentiation in *Equisetum* of Bombay Presidency including a new species (1932) from Pashan (Poona). Varadpande (1931) described *E. debile* of Nagpur. Majumdar (1926, 1930) reported *E.debile* from Pabna and suburbs of Calcutta in Bengal. Bose (1931) studied golgi bodies in *Equisetum*. Mahabale (1945) described the prothallus structure of *E. debile* var. *pashan* Bhide, an ecotype of *E. debile*, and noted some differences in the distribution of sex organs. Banerji and Majumdar (1952) studied the origin and development of the shoot apex in *E. debile*. Raizada and Chowdhury (1954, 1961) recorded five species of *Equisetum* and three varieties from the upper Gangetic plain in U.P. and its neighbourhood, i.e. *E. diffusum* Desf , *E. maximum* Lam, *E. arvense* L., *E. ramosissimum* Desf., and *E. debile* Roxb. Mahabale (1938) founded a new variety of *E. debile* Roxb. var. *pashan* Bhide, named after Prof. R. K. Bhide who was the first to observe the plant at Pashan near Poona, in 1904.

Kirtikar and Basu (1941, Vol. IV) mentioned the usefulness of *E. debile* as a cooling medicine in gonorrhoea. Sen and Nandi (1951) found the extract of *Equisetum* to be antibiotic. Amongst earlier authors, Clarke (1880) recognised only four species in India. Baker (1887), Stewart (1869), Prain (1903), Haines (1924), and Trimen (1885, 1893 & 1900) enumerated Indian and Ceylon species of *Equisetum* in their floras.

Ninan (1955) investigated the cytology of *Equisetum debile* from Mysore and Mannar; the haploid chromosome number is found to be 108. The unique feature of the occurrence of only diploids (no polyploids) with n = 108 in all the investigated species of *Equisetum* is comparable only to *Osmunda* with n = 22; both, no doubt, represent the end lines of ancient groups now at the peak of evolutionary stagnation.

Abeywickrama (1956) found *E. debile* Roxb. (rather rare) in the hilly districts of Ceylon.

Mehra and Bir (1964) reported from Darjeeling *E. debile* Roxb. (low altitudes), *E. ramosissimum* var. *attissimum* A. Br. (Tea gardens) and *E. diffusum* D. Don (exposed places, 1000-2000 m).

## VI OPHIOGLOSSACEAE

D' Almeida (1922) described six species of *Ophioglossum* of Bombay Presidency (*O. vulgatum*, *O. aitchisonii* sp. nov., *O. nudicaule*, *O. fibrosum*, *O. lucitanicum*, etc.). Saksena and Mathur (1925) gave one of the earliest accounts of an Indian *Ophioglossum*. Vashisht and Bhagat Ram (1927) compared the anatomy of the polyphyllous species (*O. aitchisonii* d' Alm.) with that of the monophyllous species (*O. vulgatum* L.) and noted the presence of scattered tracheids in the pith and di- or tri- arch root in the former, as contrasted to the monarch root and absence of medullary tracheids in the latter. Dixit and Mahabale (1934) studied the germination of spores of *Ophioglossum* of Poona. Mahabale and Deshpande (1934) studied the variability of parts in *Ophioglossum*. Mahabale (1938) described the distribution and habit of *Botrychium lanuginosum*, *Helminthostachys zeylanica*, and seven species of *Ophioglossum* of Bombay Presidency. He (1937) also described the gametophytes of *Ophioglossum* species of Poona, e.g. *O. fibrosum*, *O. pedunculosum*, *O. aitchisonii*, the fertile spike of *Ophioglossum* and dichotomy as a cause of the abnormalities in this genus. Maheshwari and Singh (1934) gave a detailed morphological account of *O. fibrosum* of U.P. The rhizome, leaf, and sporangial structures have been fully investigated; the sporangium consisting of a four-layered wall, two layers of tapetum and the inner sporogenous tissue. Govindu and Narayana (1949) described an instance of phyllody in *Botrychium virginianum.* They noted transformation of sporangia into sterile vegetative pinnules in the plants collected from Kodaikanal Hills and recalled Goebel's (1905) observations of sterile vegetative leaf being transformed to sporangia in *B. lunularia* and *B. simplex.* Chakravarty (1951) surveyed *Ophioglossum* species of India and distinguished seven species, i.e. *O. vulgatum* L., *O. reticulatum* L., *O. capense* Sw. (Syn. *O. aitchisonii* d' Almeida) *O. fibrosum* Schum., *O. gramineum* Willd., *O. nudicaule* L.f. and *O. pendulum* L. Raizada and Chowdhury (1954) recorded *O. vulgatum* L., *O. reticulatum* L. and *O. costatum* R. Br. (Syn. *O. fibrosum* Schum.) from the upper Gangetic plain. Mehta (1952) further described *O. vulgatum* L. Abraham and Ninan (1954) recorded over 1000 chromosomes at 2n in *O. reticulatum.*

L. N. Rao (1939) studied the gametophyte of *B. virginianum* var. *lanuginosum* Sw. Rao (1961) showed the presence of a definite ventral canal cell and two-neck canal nuclei in the archegonium of *Botrychium lanuginosum.* He (1961) also observed some abnormalities in *Helminthostachys zeylanica,* i.e. a bifurcating spike (cf. Banerji 1951), partial phyllody of spike, formation of axillary branch in rhizome.

Varma (1956) found the chromosome number in *Ophioglossum polyphyllum A. Br.* to be n = 116 (*O. vulgatum,* n = 256).

Majumdar (1930) reported *Helminthostachys* from the suburbs of Calcutta. Banerji (1951) noted an abnormality of forked spike and 3 segmented fronds in *Helminthostachys zeylanica* from Buxa Duars of N. Bengal. Amongst the earlier authors, Clarke (1880), Beddome (1892), Macpherson (1894), Bitter (1900), Hope (1903) and Blatter (1908) have contributed to our knowledge of this group of ferns.

Abeywickrama (1956) reported *O. pendulum* L. (epiphyte in wet zone, low country), *O. petiolatum* Hook. (terrestrial, rare), *Botrychium lanuginosum* (Wall.) Hk. et Grev., *B. dancifolium* Wall (both rare at high elevations) and *Helminthostachyas zeylanica* (L.) Hk. (very locally distributed in wet zone) from Ceylon.

Mehra and Bir (1964) found *Botrychium lanuginosum Wall.* (Syn. *B. virginianum* (L.) Sw.) to be very common in exposed places all over Darjeeling and Sikkim State (1500-2100m.), *B lunaria.* (L.) Sw. (Lachen-Sikkim); *Ophioglossum petiolatam* Hook. and *O. vulgatum* L. (Senchal-Darjeeling).

## VII MARATTIACEAE

The only account of Indian species, mostly of a taxonomic nature, is to be found in Beddome (1892) and Bitter (1900). Recently Kachroo (1953) described the occurrence of *Angiopteris evecta* from places of Assam (Gauhati, Sibsagar-Dibrugarh Road, Abor and Khasi Hills).

Blatter and d'Almeida (1922) described *Angiopteris evecta* Hoffmann from Bombay Presidency, N. Kanara and S. Konkan. Mehra and Bir (1964) also recorded *A. evecta* (Forst) Hoffm. a large-sized fern, with pinnules 1.5-3.5 cm. long, from ravines around Darjeeling and Sikkim.

C. A. Ninan (1956) investigated the cytology of *Angiopteris evecta* (n = 80), and *Marattia fraxinea* (n = 78) from S. India and concluded that the basic chromosome number in ancient vascular pteridophytes is probably 13. Abeywickrama (1956) reported *Angiopteris evecta* (Forster) Hoffm. (moist places, wet zone) and *Marattia fraxinea* Sw. (central high lands) from Ceylon.

## VIII OSMUNDACEAE

Sarbadhikari (1924, 1927) investigated the cytology of *Osmunda* and *Doodia.* He studied the somatic and gametophytic tissues and meiotic and post-meiotic mitoses in *Doodia* and *Osmunda.* He further studied (1936) the chromosome constitution of meiotic chromosomes of *O. javanica,* and the significance of chromatic bodies in it. Sarbadhikari (1937) also investigated the vascular system of *Osmunda cinnamonea.*

Sarbadhikari (1949, 50) noted aposporous development of prothalli from the margin and apex of the pinnules in *Doodia aspera.* The prothalli are formed by the activity of the marginal and submarginal cells. At first typically cordate, the prothalli during further growth form a thin transparent continuous sheet of rectangular cells. From these, the embryo arises directly as a vegetative bud. There is no alteration in the chromosome number during either process.

Chowdhury (1932) in an examination of several specimens of *Osmunda claytoniana* collected from Kashmir (Gulmarg) observed superficial abaxially placed sori on the pinnules which were like the ordinary sterile pinnules, thus the usual state of dimorphism between the sterile and fertile pinnules was not fixed in *Osmunda claytoniana.* Also, from the study of the fossil representative of the family, he concluded that at least for Osmundaceae the superficial sori was primitive in character and the marginal condition a derived state—a view directly opposite to the conclusion of F. O. Bower (1926) that the originally marginal sporangia passed to a superficial position on the abaxial surface (phyletic slide).

Bower (1935), while acknowledging the fact observed by Chowdhury (1932), unsuccessfully tried to give it a different meaning. Dickason (1946) in his phylogenetic study of the ferns of Burma completely agreed with Chowdhury's conclusions. Mehra (1932) confirmed Chowdhury's findings of superficial abaxial sori in *O. claytoniana* collected from additional localities, i.e. Almora District, Jalori Pass, Garbayang and Kashmir. Definite sori were seen on the abaxial surface; the sporangia were always raised on a massive cushion (placenta) developed either above or in between the branches of a vein. Some aborted sporangia occurred scattered between the normal ones, having massive stalks, but reduced non-functional heads. Chowdhury (1932) also found such abortive sporangia placed on linear thickenings which he thought could be placentae or fused mass of aborted sporangia.

Kachroo (1952) reported *Osmunda regalis* L. from Khasi Hills (Shillong, Mawphlong). Beddome (1883) mentioned the two species of *Osmunda* from parts of North and South India.

Mehra and Bir (1964) came across *O. claytoniana* Linn. as a high altitude fern (2400-3000m.) in Lachen valley (N. Sikkim) and Tiger Hill (Darjeeling). Fronds with dimorphic pinnae usually became fertile in May.

Abeywickrama (1956) found *O. javanica* Bl. common near streams in Nuwara Eliya (Ceylon).

## IX SCHIZAEACEAE

Beddome (1892) described species of *Lygodium* and *Schizaea* from India.

Hans Raj (1923) observed in *Lygodium*, the plants climb by leaf rachises, the leaves occurring in a single row, occasionally distichous ; the primary root bore secondary tuberous roots in the hill specimens. The stem stele is solid protostele with scalariform tracheids, the root diarch. The leaf trace is circular in cortex, at the base of the petiole it has a median protoxylem group on the abaxial side, occasionally, a small centrifugal metaxylem is also met with. In the sporangium, a three layered tapetum and small sporogenous tetrads were seen. H. R. Saini (1924) described the anatomy of sporophyte and the development of the sporangium in *Lygodium japonicum*. Kachroo (1953) reported two species of *Lygodium* (*L. japonicum*—Naga and Abor Hill, *L. flexuosm* (L.) Sw. Gauhati, Khasi and Aka Hills) from Assam. Raizada and Chowdhury (1954) reported four species of *Lygodium* (*L. flexuosum, L. japonicum, L. scandens, L. pinatifidum*) from the upper Gangetic Plain (Dehra Dun—Jhajra, Birpur, Tons valley ; Bahraich ; Pilibhit, Kheri, etc.).

Abeywickrama (1956) reported *Schizaea digitata* (L.) Sw. (open ground, wet low country), *L. flexuosum* (L.) Sw., *L. scandens* (L.) Sw. and *L. circinatum* (Burm.) Sw. from Ceylon. Mehra and Bir (1964) recorded *L. flexuosum* (L.) Sw., *L. japonicum* (Thbg.) Sw. in Darjeeling and Sikkim.

Mahabale and Kulkarni (1949) described the prothalli occurring in nature of *Lygodium flexuosum* to be cordate, symmetrical, delicate, with a 1-2 mm. thick cushion, and bearing a few club-shaped unicellular glandular hairs on the margin specially towards the anterior end. The prothallus is hermaphrodite. The structure of antheridium is normal and its dehiscence is by a single opercular cell characteristic of the family.

Later Mahabale and Kulkarni (1950) studied the prothalli of another species of *Lygodium*, *L. japonicum*, in culture and reported the structure to be identical with that of the previous species but the cushion is rather thin. No mention is made of the unicellular glands in this species. In the same note, anatomy of *L. flexuosum* and *L. japonicum* was described. The underground stem is protostelic with admixture of parenchyma, but in the aerial portion it assumes a multi-arched appearance passing through an X-shaped condition with four protoxylem groups at ends. Leaf trace is simple and pinna trace marginal.

Roy (1964) studied the cytology of *Mohria caffrorum* recording chromosome number, n=76 (bivalents and 2n= 153) from root tip.

## X GLEICHENIACEAE

Kachroo (1953) recorded *Gleichenia linearis* Clarke, from parts of Assam (Khasi, Jaintia, Aka, Abor Hills, and Jorhat). Sampat Kumaran and Gopala Iyenger (1932) worked out the life history of *Gleichenia dichotoma*. Beddome (1892) gave descriptions of *Gleichenia* species in India. Blatter and d'Alemeida (1922) described *Gleichenia linearis* Bedd. from Bombay Presidency, N. Kanara, etc. Mehra and Bir (1964) recorded *Dicranopteris linearis* (Burm.) Underwood and *Gleichenia glauca* (Thunb.) Hook. from Darjeeling and Sikkim forming thickets in drier situations at 1200 m. or above.

Abeywickrama (1956) mentioned (*Dicranopteris linearis* Burm. Gleichenia) Underwood, as very common from sea level to 7000 ft. in Ceylon.

## XI HYMENOPHYLLACEAE

Very little work has been done on Indian Hymenophyllaceae

Usha Sharma (1960) studied the morphology and anatomy of the epiphytic, rootless fern *Crepidomanes latealatum* Copeland (Syn. *Trichomanes latealatum* Christ.) collected from Coorg (Mercara). The rhizome showed a centrarch protostele with interrupted phloem. Leaf gaps absent, leaf trace collateral, sporangium (lepto sporangiate type) produced 64 tetrahedral spores (42×46 microns). Unicellular dermal hairs cover the rhizome.

Mehra and Gurdip Singh (1957) dealt with the cytology of some Indian species. Blatter and d' Almeida (1922) reported *Hymenophyllum polyanthos* Sw., *Trichomanes kurzii* Bedd., *T. bipunctatum* Poir and *T. intermarginale* Hook. et Grev. from Western Ghats, N. Kanara.

Mehra and Bir (1964) recorded *Micodium polyanthos* (Sw.) Copl. (Syn. *Trichomanes polyanthos* Sw.), *Hymenophyllum simosianum* Hook., both occurring as epiphyte or lithophyte, from the Darjeeling area. Hymenophyllaceae is widely represented by 9 genera in Ceylon, e.g. *Mecodium, Meringium, Microgonium, Crepidomanes, Micro-Trichomanes, Selenodesmium*, etc. (Abeywickrama, 1956).

## XII CYATHEACEAE

K. Biswas (1942, 1943) mentioned the occurrence and systematic position of the common tree fern *Cyathea* of the Eastern Himalayas.

Biswas (1942, 1943) considered the genus *Cyathea* to be distinct from *Alsophila* and *Hemitelia* differing in this respect from Holttum and Copeland who advocate the merging of the three genera into a comprehensive genus *Cyathea*. On the basis of his observations in the field and herbarium specimens in various standard herbaria, he concludes both *Cyathea brunoniana* (Wall.) Cl. and *Hemitelia decipiens* Scott to be forms of *C. spinulosa* Wall. On the other hand, *C. spinulosa* var. *Scotti* is suspected to be a new species. A key, distribution and detailed descriptions of these forms are given.

Ahmad, Shamsul Islam (*1952*) described the tree fern *Alsophila glabra* with stipe 16½ ft. long, about 1 ft. in diameter, and leaves about 11 1/3 ft.×6 ft. in length and breadth, as the first record from India found at Sylhet, down the slopes of small streams (Charionala, Hatkornala). Beddome (1892) had reported *Alsophila glabra* from Sylhet, P. Kachroo (1952) studied the spore germination in *Alsophila glabra*, a tree fern with an erect trunk found in the hills of Chhota Nagpur, Chittagong, Bengal and Assam. Spores collected from a tree in Mawphlong forest (Khasi Hills) germinated in Knop solution and soil producing a basal colourless cell which formed the rhizoid and an upper cell forming the prothallus. A two-sided apical cell was responsible for the further growth of the gametophyte which attained a cordate shape. Occasionally two rhizoidal cells were produced like those of *Diplazium esculentum.*

Kachroo (1953) reported *Cyathea glabra* (Bl.) Copel. from Gauhati, Mawphlong and *C. Khasyana* (Moore) Copel. from Mawphlong.

Abeywickrama (1956) recorded two endemic species of tree ferns, i.e. *Schizocaena sinuata* J. Sm., *S. hookeri* Copel. (endemic) and *Cyathea crinita* Copel. from Ceylon.

Mehra and Bir (1964) found several species of *Alsophila* and *Hemitelia* (e.g. *A. glabra, A. glauca, A. latebrosa, A. oldhami, A. ornata, A. andersonii, A. aff. Khasyana,* (a rare species) and *H. bruniana* (rare), *H. decipens* from the Darjeeling-Sikkim area. They also found *Cyathea spinulosa* Wallich as the commonest tree fern in all forests around Darjeeling and Sikkim between 1200 and 2250 m. above sea level.

Roy and Holttum (1965) made new cytological records in *Dicksonia hycronymi*, n = 130 (root tip) and *Cystedium sorbifolium*, n=56 (bivalents), 2n=112 (root tip).

## XIII MARSILIACEAE

One of the earliest contributions on Indian *Marsilia* came from S. S. Pande (1923, 1925) on the biology and life history of *Marsilia* of Lahore (*M. erosa* Willd.). He studied the very variable nature of its characters of stem, length of petiole, number and shape of leaflets, number of sori, the anatomy and development of rhizome, root, sporocarp with its sporangia and spores, and the peculiar occurrence of tubers with dictyostele and stored oil drops in its cortex. The rhizome showed the usual solenostele. A pathogen, *Albugo* (*Cystopus*), was for the first time reported by him on *Marsilia* rhizome. He also observed 3-5-6 layered root cap ; roots arising acropetally from the pericycle ; sporocarp mother cell arising from the margin of a very young leaf at 7 segments stage ; 5 to 8 macrospores as against only one matured macro or megaspore found normally) in a megasporangium ; occasionally such large spores were also seen in microsporangia. The spores, however, failed to germinate. The development up to 16 spore mother cells stage was found to be similar in both, and ultimately

64 microspores or a few (5 to 8) megaspores developed. He noted the presence of six leaflets in one specimen.

Kolhatkar (1937, 1940) studied the anatomy and life history of *Marsilia* of the Deccan (*M. poonensis* Kolh.). He found the root arising endogenously from the pericycle, an 8 to 10 layered root-cap, and the leaf initial arising from the 6th or 7th segment of the stem apex as against the 10th or 11th segment observed by Pande (1925), or 3rd or 4th segment recorded by Johnson (1898) in other species. Only one cell out of the 16 tetrads became larger and functional mega spore. No parthenogenesis took place. The segregated megaspore developed only into a ♀ prothallus with archegonium but no embryo was formed. In the ♂ prothallus four spermatogenous mitotic divisions were studied and centrospheres observed in all the 4 mitoses. Kolhatkar (1937) also found the centrosphere transformed into blepharoplast. The development of ♀ gametophyte, oogenesis and embryogeny confirmed only previous account and revealed no new facts. The rhizome, leaf, petiole, root, and the sporocarp showed usual structures. He (1940) established a new species, e.g. *M. poonensis* Kolh.

The outstanding features in this investigation are the probable $n=16$ chromosome number, 8-10 layered rootcap in contrast to the relatively thinner cap described for other species by earlier investigators, the observation that in megasporogenesis most of the 16 tetrads develop a preferred large spore but only one of these ultimately matures, the presence of a second sterile cell in the male gametophyte and the absence of any parthenogenetic behaviour as determined experimentally.

P. N. Mehra (1938) brought to notice a peculiar abnormality in *M. minuta* L. (Syn. *M. erosa* Willd.). He found that in some abnormal cases the 16-spore mother cells directly developing into 16 large irregular angular-shaped spores which were non-functional. In a few cases, in addition to these 16 spores, additional smaller spores (54-107), derived from the tapetal nuclei, were seen. Pande's (1925) observation of 5 to 8 large spores occurring with microspores was most probably a similar abnormality. Mehra (1938) noted that a plant either bore all normal type of sporocarps with typical micro and megaspores or all abnormal sporocarps with abnormal spores and sporangial conditions ; he, therefore, concluded that it might be due to a genetic disturbance in the plant and was not due to environmental factors.

G P. Majumdar (1944) studied the origin of shoot-borne roots on adult internodes of *M. quadrifolia* L. The origin of the roots could be traced by him up to a little behind the apex, and the place of the origin of the root primordium was always found between the pairs of protoxylem groups on the lower side of the stem. The stem of *M. quadrifolia* L. has a solenostele (amphiphloic) and lateral roots caused a gap in the stele.

Majumdar (1944) described the presence of up to 5 adventitious roots borne at regular intervals on the internodes of *Marsilia quadrifolia.* The roots during their emergence cause a gap in the solenostele which is soon repaired although the gap in the xylen cylinder persists longer. These roots have fixed point of origin in between the pair of protoxylem groups on the lower side of rhizome.

Mahabale and Gorji (1948) studied the sporeling and adult plants of *M. quadrifolia* L. collected from the vicinity of Ahmedabad. They also found multifoliate leaves in a specimen of *M. quadrifolia* L. collected from Amarapara (Bihar). They found the number of leaflets in an adult plant could be 3 to 8; one or both the leaflets in the lower pinna could be replaced by pair of teeth like spines, similar to those on a sporocarp, and concluded that the quadrifoliate condition is the result of the asymmetrical dichotomy of the leaf trace and that the sporocarp is foliar in nature, formed by the fusion of the distal pair of pinnae, with the lower pair of pinnae suppressed (cf. Johnson, 1898 ; Goebel, 1905 ; Eames, 1936 ; Smith, 1938).

Mahabale and Gorji (1948) noticed a succession of different types of juvenile leaves leading to the normal quadrifoliate condition in the ontogeny of the sporeling in *Marsilia quadrifolia.* The axis in the sporeling is erect and protostelic, becoming dorsiventral and solenostelic later. The early leaves are simple, bilobed, bifoliate, trifoliate and quadrifoliate in succession. In some adult individuals, a multifoliate condition from 4-8 leaflets is noticed. The authors think this a reversion to the ancestral character and argue that possibly the ancestors of *Marsilia* were forms with much larger compound leaves with many pinnae. In certain cases, the proximal pair of pinnae of the quadrifoliate leaves is altered to form one or two small spines supplied with vascular traces. This is comparable to the two teeth-like processes borne on the sporocarp and gives added support to the common belief of the foliar origin of the sporocarp.

S. C. Pandeya (1953) noted the variation in leaf form in *M. quadrifoliata.* Puri and Garg (1953) studied in detail the anatomy of the sporocarp of *M. minuta* L. and discussed its nature.

Mahabale (1949) recorded the occurrence of latex containing cells in the phloem and the inner cortical parenchyma of the petiole and stem of *Regnellidium diphyllum.* Such cells are not so far reported in ferns.

Mehra and Loyal (1959) conducted cytological investigations in 3 species of Marsilia : *M. minuta* L., *M. brachypus* A. Br., *M. quadrifolia* L , all having chromosome numbers, n= 20, indicating phylogenetic relationship with Schizaeceae (*Lygodium flexuosum*, n=60.) rather than with Cyatheaceae (*Cyathea*, n=69). *M. minuta* L. is a species complex of two cytological races, a diploid with n=20 and a triploid with 2n=60. The latter, an auto triploid, is abnormal, producing sterile monads in the sporangia (no differentiation into mega and micro sporangia). The diploid exists in three biotype forms. (*i*) Patna biotype (perfectly normal with typical mega and micro sporangia) ; (*ii*) Jullundur biotype (normal mega sporangia but micro sporangia having typical and atypical microspores) ; (*iii*) Saugor biotype (mega sporangia completely lacking, micro sporangia with a greater percentage of atypical microspores). Intra specific polyploidy is seen in *M. drummondii*, *M. hirsuta*, *M. quadrifolia* and *M. minuta.*

Gopal (1966) noticed vegetative propagation in *Marsilia minuta* L. Condensed shoots or propagules resembling dormant buds are formed under dry conditions which readily sprout when moistened. The condensed rhizome becomes dictyostelic overcrowded, overlapping leaves.

Chowdhury (1954, 1971) surveyed the *Marsilia* of India which number about a dozen species. He (1971) also observed dichotomy in a leaf of *Marsilia* from Rangoon. Raizada and Chowdhury (1954) described *M. minuta* L. from the upper Gangetic plain. Gupta (*1955*) recorded the occurrence of the African species *M. aegyptiaca* Willd. in Rajasthan (Jodhpur) where he thought it to be endemic in India.

From studies of field collections, surveys and examinations of herbarium specimens of Indian Botanical Gardens, Calcutta, Forest Research Institute, Dehra Dun and Peradynia Botanical Gardens, Ceylon, Chowdhury (1954, 1971) established a new status for *Marsilia minuta* L. var. *major* (Haines) as *Marsilia major* (Haines) Chowdhury *status novo* (Syn. *M. minuta* (L.) var. *major* Haines) on grounds of leaf, sporocarp, and size characters. His earlier similar (Chowdhury, 1949)[1] assertion about the new status of the multiple seeded rice of India, e.g. *O. plena* (Prain) Chowdhury *status novo* (Syn. *O. sativa* L. var. *plena* Prain) has since been recognised by authorities like Rev. H. Santapau (Chowdhury, 1954).

He also studied the species of *Marsilia* in India, Burma, and Ceylon and recorded about twelve species (vide distribution given in appendix II).

## XIV SALVINIACEAE

Deshpande (1943) studied the sporophytic generation of *Salvinia cucullata*—a species native to Bengal and successfully introduced in Poona. The two floating leaves at a node are determined to be dorsal and dorso-lateral in point of origin while the third dissected submerged one is ventro-lateral. The author reports that there is nothing peculiar in the anatomy of the stem and the two types of leaves, except that the internal cell of the first periclinal division of the apical segment seems to be responsible for the formation of xylem and internal phloem and not the entire solenostele of the stem. The sporocarps are observed for the first time, arising on the fertile segment of the submerged leaf, the megasporocarps being basal with microsporocarps formed above. The former contains 15-20 megasporangia while the latter from 150-200 microsporangia. Sixteen

[1]Indian Forester, **75**, 1949 : **80**, 1954.

mother cells are formed in microsporangia, which undergo usual tetrad division but the spores are usually abortive and embedded irregularly in a solid massula. Only 8 mother cells are found in megasporangium and after tetrad division only one enlarges to form a rather small megaspore with the usual three coats. The probable chromosome number is $n=12$ and $2n=24$.

Mahabale and Joyce D' Mello (1950) studied the structure of massulae in 3 species of *Salvinia: S. natans, S. auriculata* and *S. cucullata.* In *S. auriculata,* the massulae are hollow with microspores on the periphery, much in the same fashion as in *Massulites coelatus* which Sahni and Rao (1943) discovered in the Deccan inter-trappean beds and tentatively referred to *Salvinia.* In *S. natans* the pseudocellular tapetal mass remains solid, while in *S. cucullata,* the tapetal plasmodium develops few vacuoles, if at all, the tapetal nuclei persist and the microspores are haphazardly distributed. Of the three types of massulae, the hollow type is regarded as primitive.

Mahabale (1950) cleared the identity of *Sausarospermum* and *Massulites,* the two genera reported by late Prof. Sahni (1934, 1943)[1] from Deccan Inter-trappean cherts in Madhya Pradesh —the former was referred by him to represent massulae of a water fern closely akin to *Salvinia* and about the identity of the latter nothing could be stated with certainty. On the basis of comparison with the reproductive organs in various species of *Salvinia* and particularly in *S. auriculata,* the author confirms *Massulites coelatus* to represent the massulae of a species of *Salvinia* closely related to *S. auriculata* and *Sausarospermum* to be the megaspores of the same. On account of the discovery of vegetative parts of *Salvinia* in the same chert, the author suggests the two-form genera to belong to the same species which he names *Salvinia intertrappea.*

Pandit and Mulay (1931) gave one of the earlier accounts of *Azolla* of Khandala. Later Mulay (1939) gave an account of the male prothallium of *Azolla pinnata* and its micro and mega sporocarps. He (1939) described the fertilization of *Azolla pinnata* and the chromatin material (1937) found in the micro sporocarps. He (1939) also made ecological observations on its life history. Mulay (1938) further studied the

[1]Sahni, B. and Rao, H. S. Proc. Nat. Acad. Sci. India 13, 1943.

development of the female prothallium in *Azolla pinnata* R. Br. and discussed the differences and resemblances of the prothallus of *A. pinnata* with that of *A. filiculoides.*

Sud (1934) gave a preliminary account of *Azolla pinnata* R. Br. H. S. Rao (1935) too worked out its structure and life history and studied the fossil history of the Hydropterideae. He noted that in the structure and development of the stem, leaf, root, sporocarps and gametophytes, *A. pinnata* is essentially similar to other well-known species. The branching is extra axillary, an absciss layer of smaller cells is seen at the bases of the branches and root. Because of this, the plant easily fragments and the root becomes deciduous. There is a two-sided apical cell at the stem apex. In the megasporangium, only one mega spore develops and 31 of the mega spores abort. In microsporangium, 64 microspores are developed. The male prothalli of *A. pinnata* are detachable. Escaping from the massulae they come to lie about the female prothalli. The glochidia are of various shapes, being either simple processes or branched or hooked. *A. filiculoides* and *A. pinnata* are known from the pleistocene.

Only a single species of *Azolla*, i.e. *A. pinnata*, is recorded from India. Mulay (1939 a-c) suspects the occurrence of 2-3 varieties in the country and states certain divergences from different localities. He confirms previous observations of Rao (1935) that microsporocarps after detachment sink to the bottom and rudimentary glochidia are present on the massulae of the species. He confirms his own previous observations (1938) that fertilisation occurs on the soil below water.

Sen (1937) studied the effect of different cations on the living protoplasm of the root hair of *Azolla pinnata.* Ghiasuddin Ahmed (1941) studied the effect of light intensity, temperature, and mineral solutions without auxins on the growth of *Azolla filiculoides.* Raizada and Chowdhury (1954) described *A. pinnata* of U. P. Biswas and Calder (1937) described the *Azolla* and *Salvinia* species of India, and Kachroo (1953) enumerated those of Assam. Mahabale and D.' Mello (1952) observed apogamy in *Salvinia.*

Abeywickrama (1956) recorded *Azolla pinnata* R.Br. (low country), *Salvinia auriculata* Aubl. (very common, S. Western area) in Ceylon.

# XV PARKERIACEAE

Beddome (1892), Blatter and d' Almeida (1922) and Biswas and Calder (1937) had reported *Ceratopteris thallictroides* (*L.*) Brongn., a perennial amphibious fern, from various parts of India. Kachroo (*1953*) reported *C. siliquosa* (L.) Copl. from Gauhati and Aka Hills of Assam.

Mahabale (1948) described the prothalli of *C. thallictroides* Brongn., the water pool fern of India and Ceylon. The prothalli collected from Sabarmati river banks at Ahmedabad were hermaphrodite, cordate with embedded antheridia of *Anemia* type and archegonia of polypodiaceous type. No hairs of any kind existed on the prothallus. A facultative fungus *Completoria* was noticed in a prothallus. The early embryogeny was found to be normal.

The gametophyte does not throw any light on the affinities of the genus.

Later Mahabale and Javalgekar (1950) raised the prothalli of the above species in cultures. Some prothalli remained small and filiform and borc antheridia only, but others grew big, assuming the symmetrical or asymmetrical cordate outline in which the distribution of the sex organs was similar to that occurring in nature.

Javalgekar (1960) studied sporogenesis and prothallial development in *Ceratopteris thallictroides* (L.) Brongn. A large subsessile sporangium arising from a single initial produced 8 spore mother cells which give rise to 32 tetrahedral, brown, or yellow, ribbed spores of 110-125 μ size. The chromosome number was n = 78. Prothalli were ⚥ or ♂ but never ♀ in cultures. Antheridia, archegonia development was normal; androcytes 8, 32, antherozoides 2½ — 3 coiled, 24—30 μ long. Embryogeny was normal as described by Mahabale (1948).

Raizada and Chowdhury (1954) reported the occurrence of *C. thallictroides* (L.) Brongn. in swamps, ditches and rice fields of east Dehra Dun (Narkonda), Haldwani (Belarela) in U.P. It is said to be eaten as a vegetable like *Marsilia* in some parts of India.

Mehra and Bir (1964) found *C. thallictroides* (L.) Brongn. to be common at low altitudes in Sikkim and Bengal in rice swamps, tanks, ditches, floating especially during the rains.

Abeywickrama (1956) found *Ceratopteris siliquosa* (L.) Copel. common in paddy fields, ditches and marshes up to about 2000 ft. in Ceylon.

## XVI POLYPODIACEAE (Morphological)[1]
### *Davallieae—Lindsayeae*

Beddome (1892) gave one of the earliest accounts of the Indian Davalloid ferns and of *Lindsaya* and their distribution in the subcontinent. Mahabale and d'Almeida (1953) studied two species of *Davallia*.

Mehra and Bir (1964) reported *Davallia bullata* Wall. (a common epiphyte on *Ficus elastica*) around Gangtok and Andheri Khola (Sikkim), and *Davallia divaricata* Bl. (Sikkim).

Maheshwari and Bhargava (1935) described a solenostelic *Lindsaya*. P. Kachroo (1953) reported six species of *Davallia* from Assam Hills (Aka, Abor, Naga and Khasi Hills region), *Leucostegia immersa* Presl. from Shillong and Manipur and *Lindsaya cultrata* Sw. from Cherrapunji.

Mehra and Bir (1964) found *Leucostegia immersa* (Wall.) Presl. as epiphyte or lithophyte around Gangtok, Lebong forest (Darjeeling) between 1500-1600 m. They (1964), also, came across *Lindsaya cultrata* (Willd.) Sw. growing gregariously on rocks (moist situations) of Darjeeling and Sikkim between 1200-1800 m., and *L. repens* (Bory) Bedd. at lower altitudes.

Abeywickrama (1956) reported *Lindsaya* species, *Leucostegia ensifolia* J. Sm., *Davallia elegans* Sw., *D. bullata* Wall. with long, finely serrated scales, *Nephrolepis cordifolia* (L.) Pr. (high altitude) and *N. hirsutum* Pr. (low country) from Ceylon.

[1]For the sake of convenience, in general the old conception of Polypodiaceae as followed by Beddome (1892) has been retained. For more recent classification concepts in ferns, the reader is also referred to Christensen (1938) in Verdoon's Pteridology (1938) Holttum (1949) and Copeland (1947).

Holttum, R. E. : The Classification of Ferns, Biological Reviews, Camb. Phil. Soc., 24, 267-297.

Copeland, E. B. : Genera Filicum, Chronica Botanica Co., Waltham, Mass. U.S.A.

P. N. Mehra (1938) found the development of the gametophyte of *Pteris biaurita* to be of the type of *P. longifolia*, but there was no cushion formed behind the notch of the prothallus. The archegonia were eliminated but the embryo was formed by an apogamous bud. The antheridia having 18-37 mother cells producing motile sperms got dehisced by the opercular cell being bodily thrown off with some force. In the genus *Pteris* apogamy is also known to occur in *P. cretica, P. quardiaurita* var. *argyrea, P. sulcata, P. parkerii* and *P. flabellata.* Mehra (1932, 1938) noted occurrence of tracheids and reported apogamy in the prothallus of *Adiantum lunulatum* Burm. The gametophytes were raised in cultures from spores obtained from two locations, Sikkim and Kulu. The presence of fully developed tracheids in these gametophytes was found to be a normal feature in both cultures but, when these were grown submerged in water, the tracheids did not develop in such subaerial conditions. The Sikkim spores did not produce any archegonia at all, the Kulu spores reared prothallus produced apparently normal but non-functional archegonia; the embryo, however, in both cases, arose as an apogamous bud. The embryonal tissue cell contained a single nucleus, and at no stage, any doubling of nuclei was observed. Kachroo and Nayar (1952) studied the morphology of *Adiantum aethiopicum,* and (1953) gametophytes of seven species of *Adiantum* (*A caudatum, A. peruvianum, A. lunulatum, A. venustum,* etc.). The symmetrical cordate prothallus (monoecious, protandrous) was a normal feature but unfavourable environmental conditions produced asymmetrical, dichotomously branched ribbon - shaped or filamentous prothalli, and also induced vegetative propagation, i.e. secondary prothalli developing from marginal cells (*A. caudatum*). Mehra (1931) studied the gametophytes of *A. capillusveneris.*

P. N. Mehra (1944) described the rhizome anatomy of *Pteris* (*Campteria*) *wallichiana.* The rhizome is a dicyclic solenostele. The outer ring is alone responsible for giving rise to a simple c-shaped leaf-trace, the inner one taking no part. The latter, however, gives out a compensation strand which helps to close the leaf gap. The stelar system in the other investigated species of Pteridaceae is discussed.

Mehra (1949) further studied the cytology of apogamy in *Adiantum lunulatum.* The species was previously reported by him (1938) to possess obligate apogamy and showed occurrence of tracheids in the gametophyte as a normal feature. Mitotic division fails in the sporangium at the pre-meiotic stage so that only 8 instead of 16-spore mother cells are formed with double the chromosome number in the reconstituted nuclei. Then follows the normal meiosis and tetrad formation giving rise to 32 viable spores which germinate to form apogamous prothalli. The chromosome number at the first meiotic division is determined to be, $n$=88-90 (now established definitely as 90), which number continues throughout the life-cycle of the plant except at the critical pre-meiotic stage when doubling occurs. In a few cases, when 16 mother cells are formed without the intervention of a pre-meiotic failure of chromosome division, the meiosis is irregular and the spores formed are sterile. The species is considered to be a hybrid haploid.

Mitra (1919) studied the morphology of *Pteris* and *Adiantum* species. Stelar skeletons were prepared by boiling rhizomes in caustic potash solution. He noted solenostely, dictyostely and great similarity in the structure and origin of the strands supplying the leaves and side branches, thus supporting the view that "frond" is a modified shoot.

T. C. N. Singh (1929, 1938) described presence of sporangium on indusium of *Cheilanthes.* Mehra (1945, 1949) made cytological investigations of apogamy in *Adiantum lunulatum.*

### *Asplenieae*

Mehra (1949) was able to rear gametophyte of *Diplazium esculentum* up to 2 year's age. It attained a massive size of $1\frac{1}{2}'' \times 1''$ and showed repeated dichotomy of its prominent midrib. The prothalli were normally monoecious and protandrous but the smaller ones bore only antheridia, while the large ones only archegonia. Sexuality was thus quantitative and not qualitative. The antheridia dehisced by ejectment of the opercular cell and the embryo arose by normal fertilization. Mehra (1932) observed repeated dichotomy in the gametophyte of *Anisogonium esculentum.*

Mahabale and Tripti Biswas (1949) reported rhizome structure of *Diplazium* (*Anisogonium*) *esculentum* to be typically dictyostelic with a binary strap type leaf-trace characteristic of Aspleniaceae. The stolon shows an arc-shaped siphonostele with two protoxylem groups at the ends. Vegetative reproduction is by stolons and by formation of adventitious buds in the axils of leaflets, margin of pinnules and roots.

Mahabale and Venkateswaran (1953) worked out the life history of *Athyrium hohenkarianum.* Stokey (1948) described the gametophyte of *Actiniopteris australis*, from spores collected in India. The spores germinated in 5 to 7 days, the gametophyte was spatulate or fan-shaped when young but became asymmetric later on, with irregular lobes or branches, the sides tending to curl upwards forming a funnel-shaped thallus. Normal antheridia and abortive or ill-formed archegonia were seen. Apogamous embryos in the shape of small protuherances were formed in 8 to 10 days having tracheids and these ultimately developed into typical leafy plants. The gametophyte did not have the usual two-sided apical cell but developed by lateral meristem.

The author correlates the apogamy with the xerophytic nature of the plant and concludes a probable Gymnogrammoid affinity of the genus on the basis of gametophytic structure. Mehra (1944) observed polycyclic dictyostele in the rhizome of *Camptria Wallichiana.*

Mahabale and Shah (1949, 1950) studied the prothalli and anatomy of the sporophyte *Cheilanthes farinosa* Kaulf, *C. tenuifolia* Sw. and *C. albomarginata.* In *C. farinosa*, the prothalli growing in nature are large, cordate with a deep anterior notch and thick prominent midrib bearing scattered archegonia. If fertilization fails, the midrib may be prolonged to form a tuber for perrennation as in *Anogramma* and *Gymnogramma.* The prothalli of *C. tenuifolia* are relatively small but of the same pattern, although no tuber formation is reported. In *C. albomarginata* which is studied only in cultures a primary filament is formed from which three to four horizontal branches emerge which form an irregularly-lobed prothallus. The authors state there is no single a pical cell in the prothallus of this species. Comparative anatomy of the rhizome of these three species shows a range of variations from solenostele to dictyostele with 3 gaps. Leaf trace is simple Y-shaped with two protoxylem groups at the curved ends. In *C. albomarginata*, a compact 3-layered palisade is noticed which is not present in the other species.

Abeywickrama (1956) reported *Asplenium nidus* L., *Athyrium esculentum* (Retz.) Copel. (edible fronds) both in low country wet zone and *Actiniopteris australis* (L.f.) Link (dry zone, rocky places) in Ceylon.

### *Blechneae*

Mehra and Chopra (1951) studied the stelar anatomy of *Staenochlaena palustris* where the stem is found to be dictyostelic with 5-6 closely situated large meristeles and two rings of smaller peripheral bundles ("residual traces") derived from the leaf and root traces. The binary leaf trace enters the petiole with a large number of smaller peripheral bundles and root traces were given off from the base of the leaf gaps. It is emphasised that, amongst the Blechnoids, the vascular system of *S. palustris* is most complex. Kachroo (1953) reported *Blechnum, Brainea, Woodwardia* and *Stenochlaena* species from Abor, Khasi, Aka Hills and Gauhati respectively. Sarbadhikari (1924, 1927) investigated the cytology of *Doodia* (l.c. under *Osmunda*). Beddome (1883) reported species of *Blechnum, Woodwardia* from parts of India.

Mehra and Bir (1964) found *Staenochlaena palustris* (Burm.) Bedd. (a climber), *Blechnum orientale* L. and *Woodwardia radicans* (L.) Smith as extremely common ferns along Teesta Road (Darjeeling) and the last one along Lachen Road (Sikkim) also.

Abeywickrama (1956) recorded *Blechnum patersonii* (R.Br.) Mett. (dimorphic leaf) from Kandy, *B. orientale* L., and *Doodia dives* Kze., (high altitudes in shades) in Ceylon.

### *Aspidieae*

C. W. Hope (1890) described three new species of *Lastrea* from Assam, i.e. *Nephrodium* (*Lastrea*) *assamense*, *N.* (*Lastrea*) *subtriangulare*, and *N.* (*Lastrea*) *coriaceum* collected from the Garo Hills, Khasi-Jaintea Hills, and Kopili hot springs of North Cachar Hills respectively.

Mehra (1952) studied the colchicine effect on the prothallus of *Dryopteris sub-pubescens.* The spores were germinated in half strength of normal Knops solution (Chamberlain, 1935) with ·1-·5 colchicine added in sterilized petri dishes kept in diffused light and under 60-80° F. Suspected polyploid nuclei, chlorophyllous rhizoids and abnormal monstrous sperms of twins, triplets, and quadruplets were some of the reactions produced in the gametophytes. Kachroo (1953) reported eight species of *Dryopteris* and five species of *Polystichum* from Assam. Raizada and Chowdhury (1954) described ten species of *Dryopteris* and four species of *Polystichum* from the upper Gangetic plain. Amongst less known genera of Aspidiaceae, Kachroo (1953) reported species of *Diacalpe*, *Peranema*, *Egenolfia*, *Elaphoglossum* and *Ctenitis* from the hills of Assam.

Sahni (1917) while considering the vascular relations between branches and main stems of ferns studied the structure of stolon in *Nephrolepis* and regarded it as a specialised branch. He (1915, 1916) also studied the anatomy of tubers of *Nephrolepis* and anatomy and biology of *Nephrolepis volubilis*. Mahabale and Gorji (1951) worked out the sporogenesis in *Nephrolepis exaltata*. D' Almeida (1926) recorded a new species of *Nephrolepis*.

Mehra and Bir (1964) found *Nephrolepis cordifolia* (L.) Presl. growing on ground (producing tuber) or as epiphyte on *Ficus elastica*, and on tree ferns in low altitudes in the Darjeeling-Sikkim area.

Mahabale and Javalgekar (1949) studied the prothalli of three species of *Nephrolepis*. In *N. acuta* the prothalli, raised under laboratory conditions, were thin, cordate, with prominent wings and bore small hairs on the margin and upper surface, similar to those present on young leaves. The large prothalli were hermaphrodite. In *N. exaltata* and *N. paucifrondosa*, the prothalli were secured from wild state. These were also hermaphrodite but no mention is made of the presence of hairs on these.

Javalgekar and Mahabale (1959) studied spore germination and prothalli of *Nephrolepis* : *N. exaltata* Schott., and *N. acuta* Presl., their spores germinated readily and developed into adult prothalli (cordate) in 50 and 70 days respectively. Prothalli of *N. exaltata* were large, massive and deeply notched as compared to those of *N. acuta*. Branching of filamentous sporelings, vegetative propagation and regeneration by any cell of the prothallus, margin functioning like gemma and formation of small male prothalli in culture were common features in both species. Formation of antheridiophore, tracheids in the cushion region of the prothallus and gemma like cells are novel features of the genus. The trecheids were noticed in the anterior and posterior parts of the cushion in some prothalli of *N. exaltata* but not in *N. acuta* (similar tracheids have been reported in the prothalli of *Anemia*, *Adiantum* and *Lomaria*). The antheridium had 2-ring cells, a cap cell and 32 spermatocytes (*N. acuta*) ; an archegonium with a long neck of 4-5 cells high, binucleate neck canal cell, a venter canal cell and an egg (*N. exalta*)was seen.

Mahabale and Javalgekar (1959) also described the tuberiferous and endemic fern *Nephrolepis paucifrondosa* d' Almeida from Western India (Lonavla, Panchgani) growing as epiphyte on tall trees or on old walls. The protostele at the base of rhizome and tuber changes to solenostele in the mid stem (rhizome). Roots are small, fibrous, diarch. The tuber is formed from a stolon which has a polyarch stele. Sporogenesis is normal, producing 64 spores, having chromosome number, $n = 28$, as also seen in the wild varieties of *N. exaltata* Schott., and *N. cordifolia* (L.) Presl. The spores germinate in about 20 days and form delicate, light green, broadly cordate one-cell thick, hermaphrodite, about 0.4. cm. long prothalli.

### *Polypodieae* (Polypodiaceae proper)

B. K. Nayar and P. Kachroo (1953) studied the epiphytic fern *Drynaria* (*D. quercifolia*, *D. propinqua*) possessing dimorphic fronds, special humus collecting sessile scale leaves and green normal pinnatisect petiolate leaves, dictyostelic rhizome covered with peltate scales (slender paleae), non-indusate sori of mixed type, bilateral hyaline spores. The cordate, xerophytic, monoecious, protandrous, prothallus harboured an endophytic fungus. Sex organs were ventrally placed. The marginal cells of gametophyte developed mucilage on their outer walls which often got prolonged into long mucilage hairs, an adaptation helping in the conservation and absorption of moisture. Under faint light condition lateral secondary filaments were formed giving rise to fresh prothalli.

B. Sahni and S. K. Pande (1924), and Pande (1935) studied the anatomy of the xerophytic fern *Niphobolus adnascens* from the Malayan peninsula. The rhizome was dictyostelic with some five meristeles. The xerophytism was seen in the various water-retaining devices, i e. profusely branched felt like root system having velamen like cells in the outer cortex, the concertina or bellow-like extensible and collapsible water storage cells in the leaf and the peculiar peltate scales and hairs on rhizome and leaf.

Mahabale and Venkateswaran (1959) studied the morphology of rather rare polypodiaceous fern *Pleocnemia membranifolia* Presl. found in Assam (Khasi and Sikkim Hills), Mysore and Maharashtra (Mahabaleswar) States. The rhizome is erect and dictyostelic, leaves dimorphic with persistent black tapering scales over the stipes and rhizome. The sorus, reniform, situated at the end of a vein, is covered by a peltate circular indusium which ultimately splits into two and becomes kidney-shaped. The spore, 64 in number, are oval $45\,\mu \times 80\,\mu$ in size with pronounced perisporium. The six vascular bundles of the petiole are arranged in a ring at the base, but reduced to one at the apex of the leaf. Two types of prothalli were seen, a normal ⚥ cordate and other irregularly-shaped ameristic ♂, both having prominent hairs. The embryo developed a massive foot. This fern has probably arisen as a mutant from *Aspidium*. It also occurs in North Australia, Malaya, Southern China, Java and Ceylon.

Mehra (1952) studied the colchicine effect on the prothallus of *Goniopteris prolifera* and produced abnormal spermatozoids. He noted stout polyploid filaments, large and giant sperms some of them double-headed, large twins or triplets. The prothalli producing giant (C types) sperms never grew beyond a filamentous or at best 3 cells wide stage and bore gigantic antheridia with prominent nipple-like bulged opercular cell. The cytology of the production of these monstrous sperms, however, remained unknown. Cheema (1919) studied the rhizome and frond of *G. prolifera.*

Kashyap and Mehra (1934) described dichotomous branching in the leaves of *Pleopeltis simplex* Sw. collected from Jumna valley (Tehri Garhwal). The normal leaf was found to be simple, lanceolate, entire, but fully developed fertile fronds bifurcated once or twice.

A. R. Rao (1946, 1949) described the sporophyte anatomy and the prothallus of *Hemionitis arifolia* Sw., a xerophytic fern of polypodiaceae, found at slight elevations in South India. The rhizome shows a dictyostele with 2-3 meristeles. Both the pith and cortex are sclerised. The leaf trace is simple C-shaped and continues like that in the basal region of the petiole, but upwards, assumes an unequal X-shaped form and finally splits into 3 strands on entering the lamina. The leaf is dorsiventral with poor development of palisade but the spongy tissue is massive with considerable aeriferous region between the stellate cells. The prothallus is cordate with a ventral cushion and bears normal type of antheridia. No mycorrhiza is present. The prothalli, gregarious dark green, cordate, bilobed 3 to 4 mm. broad, devoid of any mycorrhiza, were found mixed with moss protonema. Mehra (1953), Mehra and Bir (1964) reported for the first time *Pityrogramma calomelanos* (*L.*) Link. (*Ceropteris calomelanos* L.) from India as occurring wild in Sikkim. The species had so far been found in Tropical America, Africa and Natal. Mahabale (1947) described the prothallus of *Gymnopteris variabilis* var. *lanceolata.* T. C. N. Singh (1928, 1931) described the anatomy of some polypodiaceous ferns of Mussoorie and Naini Tal. Majumdar (1933) described the morphology, anatomy, and vegetative reproduction in *Polypodium proliferum* Roxb.

Mahabale (1947) studied the prothallus of *Gymnopteris variabilis* var. *lanceolata* and found the prothallus to be dark green,

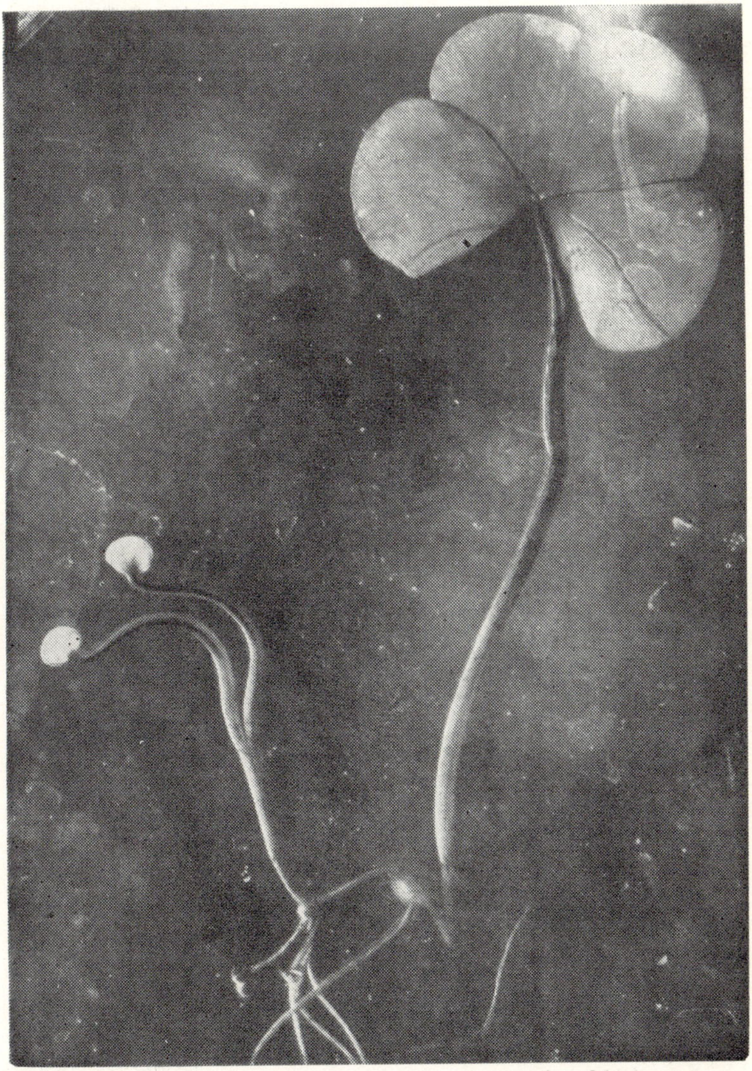

Plate I Leaf dichotomy in a *Marsilia* from Burma.

Collection : Rangoon<br>Courtesy of Prof. L. P. Khanna<br>Rangoon University

Photo and observation<br>N. P. Chowdhury.<br>(natural size)

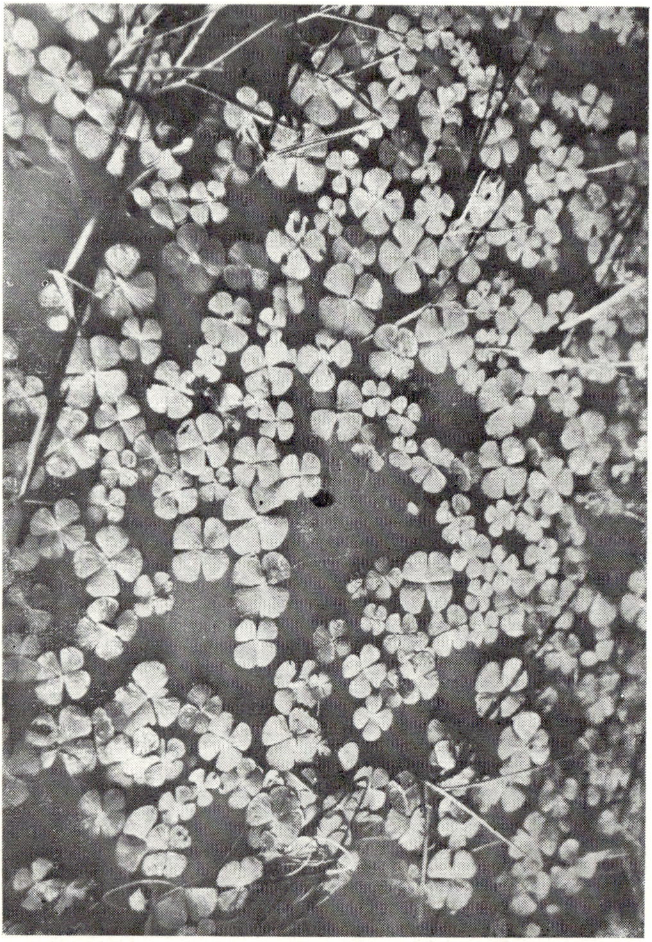

Plate II *Marsilia major* (Haines) Chowdhury status novo, flourishing in the natural habitat (canal side) near Lucknow.

Photo : N. P. Chowdhury (1/3 natural size)

Plate III Juvenile plants of *Marsilia major* (Haines) Chowdhury status novo raised by the author in the Botanical Research Laboratory of Lucknow University.

Photo : N. P. Chowdhury (natural size)

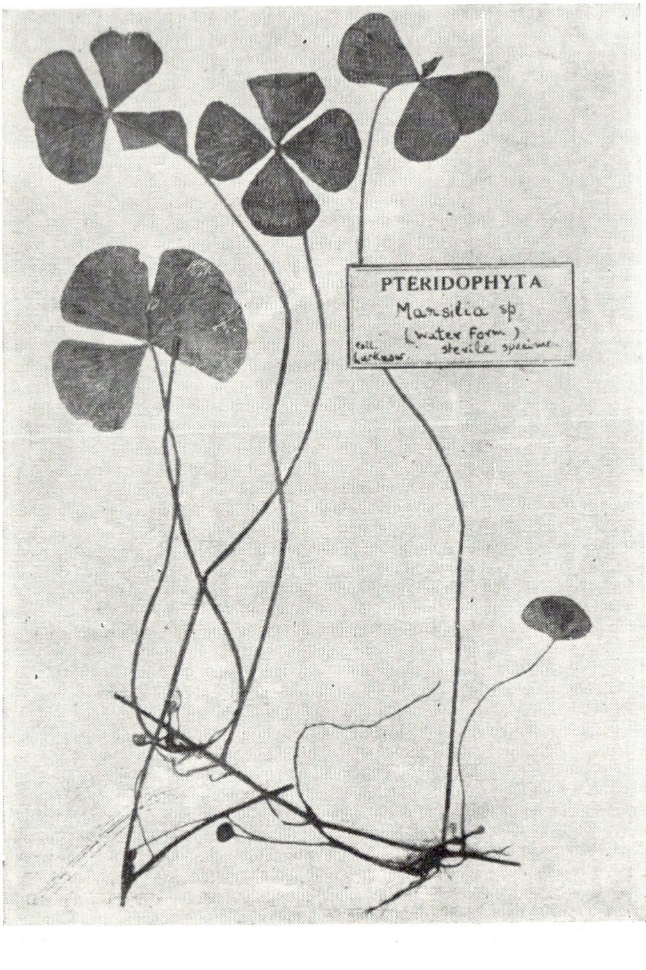


Plate IV *Marsilia major* (Haines) Chowdhury status novo.
Vegetative specimen.

Collection : Lucknow by N. P. Chowdhury

Photo : Kew Herbarium, Surrey, England (natural size)

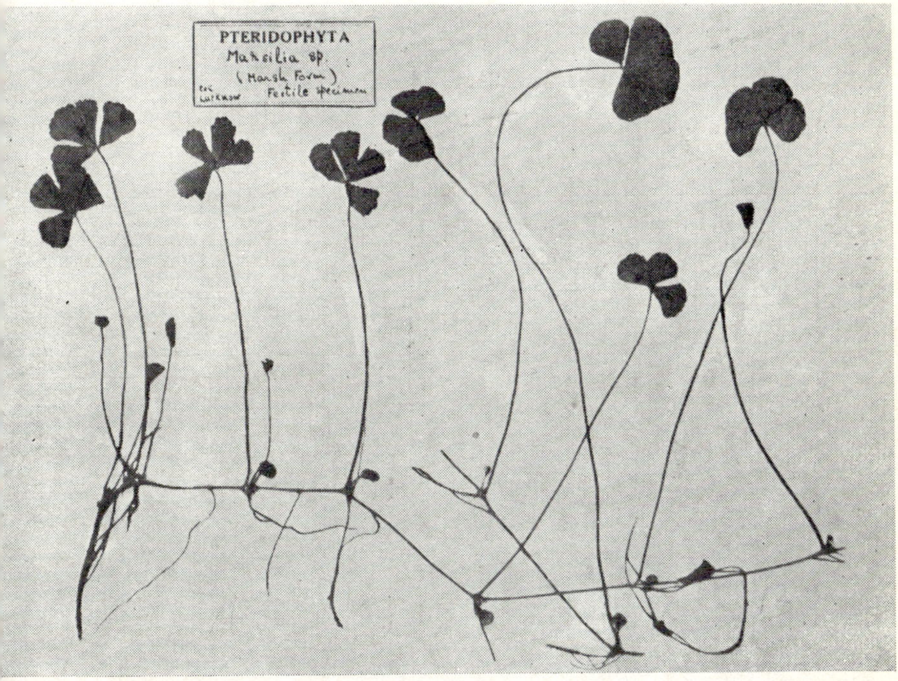


Plate V *Marsilia major* (Haines) Chowdhury status novo. Fertile specimen. Long petioled leaf bearing obcuneate leaflets and with generally solitary, non-ribbed, spatulate (bean-shaped) sporocarps.

Collection : Lucknow by N. P. Chowdhury

Photo: Kew Herbarium, Surrey, England (natural size)

spathulate, linear or irregular in outline, growing by a marginal meristem and lacking a mid-rib. Marginal gemmae are present which may not detach but grow out to give an irregular contour to the prothallus. The sex organs are typically leptosporangiate. The archegonia are borne on 2-3 cell thick tissue which may not be regular in distribution.

Abeywickrama (1956) reported *Pleopeltis thunbergiana* Kaulf. (epiphyte or lithophyte—Nuwara Eliya), *Pyrrosia zeylcynica* Geism. (endemic), *P. adnascens* (Sw.) Ching (on rocks) from Ceylon.

## XVII FILICALES

(*General : Cytology, Ecology, Palynology, Morphogenesis, etc.*)

Sahni (1917) discussed the evolution of branching in Filicales. He concluded that, in general, in Filicales the branches do not hold any regular position with respect to the leaves. The fact that the stele of the so-called axillary branch in Hymenophyllaceae and Ophioglossaceae is sometimes confluent with the leaf trace of the subtending leaf shows that the 'undivided trace' is of a dual nature—a view originally advanced by Scott, but later rejected by him in favour of the foliar view. Also, the branching in ferns ranges between two extreme types, one in which the rhizome divides into two more or less equal branches and the other in which the proximal part of one of the branches attains a great length and bears either reduced leaves (*Struthiopteris*) or none at all (*Nephrolepis*). This portion is termed 'stolon' which in extreme cases is capable of producing a large number of lateral branches each of which is a potential individual. The most important direction in which the filician branch system has progressed is specialization for vegetative propagation and subsidiarily for epiphytism and food or water storage. Sahni (1925) also discussed the ontogeny of vascular plants in the light of recapitulation which phenomenon is of wide occurrence in plants, and, in this connection, he cited interesting observations in *Tensipteris*, *Lycopodium*, *Selaginella*, *Osmunda*, *Cibotium*, *Alsophila*, etc. The occurrence of scattered xylem elements in the pith of a wounded *Osmunda regalis* is an instance of ( ? traumatic) reversion to juvenile condition ; the juvenile leaves of *Osmunda* possessing a solid mesarch strand while the later leaves possessing endarch trace ; the leaf trace of *Cibotium barometz* being undivided, C-shaped at its origin, but distally getting divided into a broken arch ; and the 'Lindsaya' phase of stem-steles, are all instances of the phenomenon of recapitulation in ferns.

Ghias-ud-Din Ahmed (1941a) made physio-ecological observations on the growth of *Azolla filiculoides* in relation to temperature and light. The cultures were made in Hoagland solution without any organic matter extract. Light and temperature were found to act as limiting factors, as increase in any one of them produced greater growth. In low intensity of light, the growth in the root system was poor which rapidly improved on increase in illumination.

In another experiment (Ahmed, 1941b) when the species was grown in simple Hoagland solution without any organic extract, it grew continuously for four months without showing signs of deterioration under normal conditions of light and temperature. The author concludes that probably the plant is able to manufacture 'auximone' within itself. When, to Hoagland's solution, extracts of peat and yeast were added, the growth was enhanced by 8-10% in one week.

Cytological investigations by Abraham, Ninan and Mathew (1962) broadly revealed : (i) species complexes in *Adiantum*, *Thalypteris pseudocalcrata* and *Asplenium unilaterale*—a new cytological type, n=116 (40×2+36); (ii) presence of intergeneric and species hybrids, i.e. *Tectaria decurrens* (*Tectaria* × *Quercifilix*) and *Adiantum calopodes*, *Blechnum orientale*, *Phlebodium aureum*, *Pteris ensiformis*, *Asplenium aethiopieum*, etc., showing irregular chromosome pairing suggestive of species hybrids; (iii) phenomena of polyploidy (auto and allello), aneuroploidy in *Adiantum* (*A. cuneatum*, n=37 ; *A. hispidum*, n=171), *Ceratopteris siliquosa* (n=40, n=77) and apogamy in *Pteris biaurita*, *Hemionitis arifolia*, *Cyrtomium falcatum*, *Adiantum caudatum*, etc. ; (iv) no correlation exists between chromosome size and chromosome number (like higher the number smaller the size or vice versa) as exemplified by (a) *Marattia fraxinea* (1.4—4.4 μ. size, 2n=

156), *Ophioglossum reticulatum* (1.5—4.5 μ. size, 2n=1260), *Selaginella* sp. (1.5—4.5 μ. size 2n=18), (b) *Isoetes coromandelina* (2.0—4.0μ, 2n=22, 3.3—6.0μ, 2n=33), *Isoetes sampatkumarani* (2.2—4.0μ, 2n=66), *Angiopteris evecta* (2.2—5.0μ, 2n=160), (c) *Psilotum nudum* (4.5—12μ, 2n=208), *Osmunda regalis* (6.6—10μ, 2n=44).

Mehra (1961) exhaustively dealt with the species complexes, polyploid evolution in tropical and temperate ferns, apomixis and impact of cytological investigation on the understanding of fern phylogeny.

Manton and Sledge (1954) studied cytology of pteridophyte flora of Ceylon.

Sinha (1929) discussed the origin of archegonium and its structural elaboration in the Archegoniatae. According to him, the archegonium had its origin in Charales (Thallophyta).

Mahabale (1945) dealing on the concept of vestigial organs and their importance in vascular cryptogams mentioned the vestigial traces of xylem found in the teeth of the sporocarps of *Marsilia* and *Pilularia*, in the ochreola found at the bases of the branches of *Equisetum* species (*E. arvense*, *E. limosum* and *E. hienale*). The cambium found in the stem of *Botrychium*, *Equisetum maximum*, and *Isoetes* is also thought to be of vestigial nature.

He first considered the characteristics of the vestigial organs and then classified them into four categories, embryonic organs, rudimentary organs, arrested organs and vestigial tissues. The topic is discussed in a comprehensive but generalised fashion. The author suggests that many of these vestigial organs may have had their origin as a result of gene mutations which, having been found useless, may have been still further reduced due to natural selection.

Mahabale (1963) discussed the evolutionary tendencies in the genus *Azolla* which is essentially a Tertiary genus. Three major complexes in *Azolla* are noticed, but the chromosome number remains more or less constant (n=8) in all the species. Apogamy played a considerable role in the evolution of the genus as in *Salvinia* (Mahabale and d' Mello, 1952).

Main evolutionary trends are : (i) Stele character viz. protostele in A. *filiculoides* complex and *A. pinnata* ; amphiphloic siphonostele in *A. nilotica* ; (ii) Massulae and glochidia conditions, viz. glochidia long anchor-shaped with many partitions in *A. filiculoides*, *A. maxicana* ; glochidia anchor-shaped with few partitions or none in *A. rubra*, *A. prisca*, *A. intertrappea*; blade of anchor lost, massulae hairy in *A. pinnata* and, finally, both blade and stem of anchor lost, massulae eglochiate as in *A. nilotica.*

Majumdar (1937) described the morphological nature of the compound leaves of ferns. Singh (1931, 1932) described forking of fronds, ventilating system and origin of double indusium in some Indian ferns.

Agharkar (1938) enumerated some important pteridophytic researches of India in Beni Prasad's Progress of Science in India (1938).

Mehra and Bir (1964) gave an excellent account of the ecology and distribution of the ferns and fern allies of Darjeeling-Sikkim region. Some of the xerophytes, lithophytes and higher altitude lithophytes named are *Cheilanthes tenuifolia* (Burm.) Sw. *Onychium siliculosum* (Desv.) C. Chr., *Elaphoglossum conforme* (Sw.) Schott., *Vittaria sikkimense* Kuhn., *V. ophiopogonoides* Ching, *Leucostegia immersa* (Wall.) Presl., *Microsorium zippelii* (Bl.) Ching., *Asplenium cheilosorum* Kunze, *A. griffithianum* Hook., *A. tenuifolium* Don., *Polystichum lentum* (Don.) Moore, *Denstaedtia appendiculata* (Wall.) J. Smith, *Lycopodium clavatum* L., *Selaginella caulescens* Spr., *Athyrium duthiei* (Bedd.) Bedd. (4,200m.) and *Cystopteris sikkimense* Ching ex Bir (4000m.).

Sen and Nandi (1951) extracted antibiotics from the Pteridophytes and found the antibacterial property to be greater in Polypodiaceae and negligible in water ferns (*Marsilia*). *Drynaria* and *Adiantum peruvianum* were regarded by them as hopeful antibiotics ; observations on *Lygodium*, *Nephrolepis*, and *Polypodium* species were also made. Handa, Kapoor and Chopra (1947) assayed the anthelmintic property of *Dryopteris* species ( male ferns ) of Kashmir. *D. odontoloma* (Pahalgam), *D. marginata* (Tan marg) of Kashmir contained 2.8 and 3.03 per cent Filicin and were up to B. P. and U. S. P. standards based on *D. filixmas* of Europe and *D. marginalis* of the U. S. A , both not indigenous to India.

Mehra and Mittal (1961) found that "filicin" content does not increase with the increase of chromosome number in the natural tetraploides of the medicinal fern *Dryopteris filixmas* which produces oleoresin and "filicin" secreted by internal glands in the schizogenus cavities.

Mahabale (1962) mentioned the occurrence of buds (vegetative reproductive organ) in Indian ferns. Buds are known to occur in some 219 species of ferns and arise variously on the fronds : margin of leaflet (*Goniopteris subcrenata*); on stipules (*Angiopteris*) ; axils of pinnae (*Cystopteris bulbifera*); base of leaf or pinnules (*Asplenium bulbiferum*, *Anisogoium esculentum*) ; end of rachis (*Anemia*, *Adiantum*, *Athyrium*, etc.) ; also axillary (*Botrychum*) or on roots (*Ophioglossum*, *Botrychium*, *Platycerium*. etc.). In *Ceratopteris*, they occur on the margin, base or apex of leaf. In *Goniopteris prolifera* Roxb. they develop on the axils of leaflets borne on 7-8 ft. long flagellae (rachis). The buds here produce either vegetative or sporophylous shoots (dimorphie). Application of I.A A. inhibited growth of roots in buds. Chaemographic analysis showed presence of amino acids (cystine, phenyl alamine, proline, threonine and lencine) in the young buds. They also contain valine, methionine, isolencine glucose and sucrose. The occurrence of two types of buds is probably due to some hormone produced at the tip of the rachis.

Startling (1888) was perhaps the earliest in India to culture fern prothalli. He successfully germinated fern spores to produce mature plants within 15 months or so. The spores gathered from fronds were sprinkled lightly on a fine layer of burnt brick powder overlying a pressed layer of fine charcoal. The spores were finally covered with a glass. Water gently poured on the outer surface of the glass just moistened the earth in the pot. After a time, bright green prothalli appeared. Young plants when bearing three fronds, were carefully taken out and planted in shade.

Saha (1963) carried out morphogenetic studies on the distribution and activities of leaf meristems in ferns. Wardlaw and Mitra (1958) similarly studied the responses of shoot apices of *Dryopteris aristata* and 'detached' meristems on the rhizomes of *Onoclea sensibilis*, *Matteuccia struthiopteris* when treated with various concentrations of naphthalene, acetic acid (N.A.A.), gibberelic acid (G.A.) and kinetin. Meristems treated with N.A.A. developed into arrested buds (*Matteuccia*) or reduced plantings (*Onoclea*) or stunted plantings with deformed leaves (*Onoclea* with Kinetin). In apices of *Dyopteris*, N.A.A. caused arrest of activities of apical meristem. G.A. caused formation of double primordia ; Kinetin (100 mg/1) proved lethal (.1 mg/1) slightly stimulatory to general growth.

Raghavan (1964) found the analogs treatment (purines, adenine, guanine, pyrinidine, cytosine, uracil) completely inhibited the two dimensional differentiation in the gametophyte of *Asplinium nidus* L.

The study of pteridophyte spores is a modern trend likely to serve as a morphological entity and criterion in the classification of pteridophytes. An important contribution in this field of Palynology is by B. K. Nayar (1964) and P. N. Mehra (1968).

Nayar (1964) described spore characters, i.e. shapes, size, exine, perine features in different families. The different shapes met with are :

I BILATERAL, MONOLETE : Psilotaceae (*Psilotum*, *Tmesipteris*—plane-convex to concavo-convex) ; Davalliaceae (*Davallia*, *Leucostegia*, *Nephrolepis*) ; Blechnaceae (*Staenochlaena palustris*) ; Aspleniaceae (*Asplenium*, *Diellia*) ; Aspidiaceae (*Elaphoglossum*, *Polystichum*, *Athyrium*).

II GLOBOSE (SPHERICAL) : Equisetaceae ; Osmundaceae (both globose, tetrahedral) Cheilanthoid ferns (*Cheillanthes chrysophylla*).

III TETRAHEDRAL, TRILETE : Lycopodiaceae ; Selaginellaceae ; Ophioglossaceae ; Loxomaceae, Hymenophyllaceae ; Parkeriaceae ; Cyatheaceae ; Matoniaceae ; Marsiliaceae ; Salviniaceae.

IV BOTH BILATERAL, TETRAHEDRAL : Marattiaceae (bilateral monolete in *Christensenia*, *Danaea*, *Marattia*—both types; tetrahedral trilete in *Angiopteris*, *Archangiopteris*, *Macroglossum*) ; Pteridaceae (bilateral in *Lindsaya*, *Hypolepis*, *Paesia*, *Lepidocaulon* ; tetrahedral in most species, *Lepidolepia*—both types); Polypodiaceae (mostly bilateral, monolete ; tetrahedral trilete in *Holttumiella* of Borneo and *Chaenopleura* of New Guinea); Vittariaceae (bilateral in *Vittaria*) ; Schizaeaceae (bilateral monolete in *Anemia, Lygodium, Mohria* ; tetrahedral, trilete in *Schizaea*).

## XVIII PTERIDOPHYTA

### (*Taxonomic*)

The fioristic accounts of the Pteridophytic vegetation of the subcontinent have been given from time to time by many authors for various localities. Thus, to name a few of them as random examples, the following could be mentioned : Clarke (1880) Ferns of Northern India ; Beddome (1883, 1892) Ferns of S. India, British India, and Hand book to Ferns of India ; Baker (1887) Fern allies of India ; Alston (1945) *Selaginella* species of India and Pakistan ; Hope (1899) Ferns of North-West India ; Singh (1928, 1931) Ferns of Nainital and Mussoorie; Mahabale (1937) *Equisetum* species of Bombay ; Chakravarty (1951) Indian *Ophioglossum* ; Bharadwaja (1935) *Isoetes* in India; Raizada (1935) *Psilotum* in India; Mehra (1939) Ferns of Mussoorie ; Macpherson (1894) List of Ferns of N. Kanara ; Stewart (1938, 1939, 1945, 1951) Ferns of Dehra Dun, Mussoorie, Kashmir and Pahalgam respectively ; Blatter (1909) Ferns of Bombay Presidency; Raizada (1952) Ferns of Nepal; Kachroo (1953) Ferns of Assam ; Raizada, Chowdhury (1954) Ferns and Fern allies of the upper Gangetic plain ; Majumdar (1933) Fern flora of Calcutta and its suburbs ; Haines (1924), Mooney (1950), and Prain (1903) Pteridophytes of Bihar, Orissa, and Bengal respectively in their floras. Mehra and Bir (1964) described the pteridophytic flora and ecology of Darjeeling and Sikkim. Similarly, Raizada and Chowdhury (1961) described Fern allies of the upper Gangetic plain.

Chowdhury (1937) *Lycopodium* species of India, Burma and Ceylon; Biswas (1935) Ferns and Fern allies of Burma; Dickason (1946) Ferns of Burma ; Blatter (1908) Ceylon Ferns; Blatter and d'Almeida (1922) The Ferns of Bombay; B.A. Abeywickrama (1956) The genera of Ceylon pteridophytes; and R. R. Stewart (1957) The Ferns and Fern allies of West Pakistan and Kashmir, are some others to mention.

Alston (1945) enumerated 57 species of Indian *Selaginella* and gave their authentic distribution and nomenclature.

Panigrahi and Dixit (1966) gave an account of taxonomy, ecology and distribution of 24 species of *Selaginella* of India.

Panigrahi and Chowdhury (1962) gave an enumeration and distribution of fern allies in eastern India.

Biswas (1939) described two new species of *Elaphoglossum*, *E. ballardianum* Biswas and *E. krajinae* Biswas from S. India. The former is distributed in the Nilgiri Hills ascending up to 2438m altitude while the latter has the same area of distribution and also Pulney Hills, and high hills of Ceylon, ascending up to 2438 m. The systematic position of the genus is discussed.

Biswas (1942, 1943) recorded the genus *Cyathea* as distinct from *Alsophila* and *Hemitelia* differing in this respect from Holttum and Copeland who advocate the submergence of the three genera into a comprehensive genus *Cyathea*. On the basis of his observations in the field and herbarium specimens in various standard herbaria, he concludes both *Cyathea brunoniana* (Wall.) Cl. and *Hemitelia decipiens* Scott to be the forms of *C. spinulosa* Wall. On the other hand, *C. spinulosa* var. *Scotti* is suspected to be a new species. A key, distribution, and detailed descriptions of these forms are given by him.

Panigrahi and Patnaik (1968) gave an account of the distribution and ecology of the Polypodiaceae in Eastern India. Most of the species are epiphytes under the canopy of tropical, subtropical and temperate forests. Some dominant genera recorded are : *Pyrrosia* (20 spp.), *Pleopeltis* (17 spp.), *Microsorium* (13 spp.), *Polypodium* (8 spp.); a few are monotypic, i.e., *Dipteris*, *Platycerium*, *Xyphopteris*, *Grammitis*, etc. seven species are endemic to Assam and E. Himalayas, viz. *Polypodium atkinsonii*, *Dipteris wallichii*, *Pyrrosia mannii*, *P. stenophylla*, *P. subvelutina*, *Leptocheilus minutulus* and *Drynaria meeboldii*.

Mehra (1939) published an account of ferns of Mussoorie as a Punjab University Lahore,. publication. A total of 99 species including the fern allies were listed with copious notes regarding their distribution and ecological habitats illustrated with photographs. Bower's system of classification and nomenclature by Carl Christensen were followed. One new species *Polypodium kashyappii*, and two new varieties of *Polypodium microrhizoma* were discovered.

Stewart (1942) gave a systematic account of fern and fern allies of Mussoorie and Dehra Dun within a range of 2,200 to 7,500 ft., listing a total of 112 species. The collections were made in seven summers from 1930 onwards. Species reported by previous workers and not observed by the author were also mentioned. Christensen's nomenclature was followed.

Stewart (1945) gave an excellent account of ferns of Kashmir based on his personal collections and those of his colleagues and earlier workers. The area is not reported to be so very rich is ferns, particularly the part behind the main Himalayan range where the only fern is *Cystopteris fragilis.* A total of 106 species and varieties are listed besides the 7 species of fern allies. Brief notes are given and localities of distribution mentioned. Carl Christensen's nomenclature is adopted. A comparison is made with the ferns of the areas west of Kashmir. A point of interest brought out is with regard to the distribution of *Microlepia wilfordii* Moore which is not so far known in India, but which grows in Japan, Korea and China.

Dickason (1946) in the first part of his paper on 'A phylogenetic study of the Ferns of Burma' gives a brief review of the schemes of classification of ferns proposed since the beginning of this century, and then discusses the various morphological concepts and the physiological factors that form phyletic patterns of variation of the Leptosporangiate ferns in a most unorthodox fashion. In the second part, 400 species belonging to 104 genera are listed. Dickason agreed with Chowdhury's (1932) conclusion regarding sorus position in Osmundaceae. Carl Christensen's nomenclature, with additions and alterations from the works of others, chiefly Ching, is followed. There is a combination of elements from the China-Himalayan region and the Malay Peninsula.

Stewart (1957) gave an account of some 127 ferns and 10 fern allies of W. Pakistan and Kashmir. The fern allies included *Lycopodium selago* L., *Selaginella rupestris* (L.) Spring, *S. sanguinolenta* (L.) Spr., *S. chrysorhizos* (Hk. and

Grev.) Spr., *S. chrysocaulos* (Hk. and Grev.) Spr., *Equisetum arvense* L., *E. diffusum* Don, *E. palustre* L., *E. debile* Roxb., and *E. ramosissimum* Desf.

Abeywickrama (1956) published a comprehensive account of the pteridophytes genera of Ceylon giving systematics, distribution and ecology of some 98 genera of the ferns and fern allies. The endemic species include *Schizaea* 2 species — *S. sinuata* J. Sm., *S. hookeri* Copel.; *Bolbitis* 3 species —*B. lanceolata* Hook; *Pyrrosia zeylanica* Hk.; *Selaginella omithopodiodes* (L.) Spring, and some of the rare species are *Psilotum nudum* (L.) Beauv., *Actiniopteris australis* (L.) Link. (dry zone rocky places), and *Acrostichum aureum* L. (mostly in brackish water). Some earlier contributions to the pteridology of Ceylon are from Fergusson 1886 , Wall 1873, and Willis 1911.

# APPENDIX I

Some current contributions, which became available when the book was almost at the final stage of print, are appended here.

Pant and Khare (1969) described the stomatal ontogeny and the structure of the mature epidermis of *Tmesipteris tannensis* Burnh. The development of the stomata is perigenous like that of *Psilotum.* The epidermal cells show sinuous outlines in surface view and have alternating lignified and unlignified transverse bands. The guard cells are sunken and have similar-lignified bands on their surface (gymnosperm type). The cuticle showed the cell and stomata outline. Lignified lamellae are also seen over the guard cells of *Psilotum*, but are absent over its epidermal cells. *Tmesipteris* is thought to be unique in vascular plants in having lignin bands over its epidermal cells.

Jyotsna Chatterji and Mohan Ram (1968) studied the gametophytes of *Equsetum ramossissimum* Desf. collected from the eastern bank of Jamuna river in Delhi and also as grown in culture (Moore's medium). The prothalli were ♂, ♀ or ⚥ reddish brown, compact, cushion-like, appearing from November to February in nature, but were flimsy, erect and much branched in culture. Sex organs got differentiated in the peripheral meristem. Antheridium with a single-layered jacket contained numerous multiciliate, spirally coiled spermatozoids escaping through a pore caused by the separation of the cover cells. Each androgonal cell divided mitotically to form two spermatocytes which developed individually into a spermatozoid. Archegonium had 2 equal or unequal neck canal cells, one or, occasionally, two ventral canal cells and an egg. Presence of a thin nuclear membrane around the egg nucleus and the spermatozoids casting off the blepharoplastic band bearing cilia when penetrating the egg cytoplasm are special observations of the authors (ct. Sethi, 1928). Numerous archegonia get fertilized, but only a few mature into sporophytes.

Sunanda Pal (1969) studied homology in the development of dermal appendages and soral structures in Polypodiaceae. The dermal appendages, like trichomes, scales and paraphyses on the sporophyte, and the trichomes, rhizoids and marginal hairs on the gametophyte have been compared in their development.

Bhambi and Prakasa Rao (1969) observed that root apex organization in pteridophytes is complex with many variations. In leptosporangiate ferns, the roots are adventitious, arising endogenously in the rhizome at the base of very young leaf primordia. The root apex has a large pyramidal apical cell with 4 cutting faces, dense cytoplasm and a prominent nucleus, but becomes vacuolated and prismatic later on. In the eusporangiate fern, *Angiopteris evecta*, roots are adventitious, root apex consisting of a group of centrally placed initials of equal-sized large rectangular cells which also become vacuolated and prismatic afterwards. In the fern allies (*Isoetes coromandelina, I. panchananii*), roots are adventitious, with root apexes having a group of centrally placed cells (root initials). The mature root structure is, however, very varied, i.e. monarch in *Isoetes* diarch in *Actiniopteris,* di-or tri-arch in *Lygodium* and polyarch in *Angiopteris*. In *Marsilea strigosa,* the root apical cell is a 'unicellular quiescent centre.'

Panigrahi (1969) thinks that recent studies in the systematics and cytology of *Selaginella*, a natural genus with about 800 species and of the epiphytic family Polypodiaceae indicate the fact of prolific evolution at species level in *Selaginella* and at generic and species levels in the family Polypodiaceae sensu stricto.

Verma (1969) studied the genetics of gametophytic variation and mating system in *Adiantum capillus veneris*. He observed that inter gametophytic selfing reduced the level of heterozygosity whereas intra gametophytic selfing released homozygotes, and both these factors have important bearing on the polymorphicity and habitat variation of the species.

Verma and Molina Goloknath (1969) studied the morphology and breeding system of *Cheilanthes tenuifolia* (Burm.) Sw. They noted that there is a general uniformity in the external morphology of this species collected from various localities but the spore morphology of the Himalayan (Darjeeling) sample is different from the previous reports from New Zealand, Australia and India. The spores are coffee-coloured, tetrahedral subglobose and trilete. Amb is triangular (when tetrahedral) with broad round corners, and average equatorial diameter of 53 microns. Perine closely adheres to exine. The Ceylon and South Indian samples are known to be sexual tetraploid but the E. Himalayan material is found to be diploid apogamous with n=56. The archegonia were non-functional.

Sharma and Tripathi (1969) described the ontogeny of leaf buds in *Adiantum caudatum* L. ; the bud primordium was seen as an adaxial raised tissue mound covered with dark grey brown scales and hairs at the extreme tip of the frond. The bud apex comprises a superficial layer of narrow prismatic cells with a median 3-sided apical cell. Endogenous roots precede the formation of leaves on the bud, the root initial cell is 4-sided, quadri seriate, densely protoplasmic with a prominent nucleus

Mehra and Asha Gupta (1969) studied the gametophytes of 4 species of *Onychium* (*O. auratum, O. lucidum, O. plumosum O. contiguum*) and 2 species of *Cryptogramma* (*C. crispa* and *C. stellari*). The spores of *Onychium* species are large non-perinous and possess 3 prominent equatorial bands of thickening as in *Anemia* and *Mohria* (Schizaeaceae). *Cryptogramma* spores were not so patterned. The mature gametophytes are cordate but in *O contiguum* highly asymmetrical with wings variously lobed, re-curved and ruffled. In *O. lucidum*, collenchymatous thickenings are present in old prothalli Twin-celled projections occur along the margin in all the prothalli seen in the gametophytes of *Mohria* and *Anemia*. The antheridia in *Onychium* are of the leptosporangiate type. In *O. auratum*, however, a primitive type exists, i.e. with 2 basal cells, 2 ring cells or 2 opercular cells. The archegonia developed along the margin, showing double axial rows of cells in the neck, or two neck canal cells instead of two free nuclei. The spore sculpture (of *Onychium*), absence of apical cell, appearance of lateral meristem in young prothalli, twin-celled projections from the margin, primitive type of antheridia (in *Onychium*) and the presence of multicellular hair in *C. crispa* establish a firm affinity between *Mohria*, *Anemia*, *Onychium* and *Cryptogramma* and the derivation of the family Cryptogrammaceae from the schizaeaceous stock.

Bir (1969a) in his study of the role of anatomy in the taxonomy of asplenioid and athyrioid ferns observed that, anatomically, the members of these groups differ markedly from each other. Asplenioid ferns are characterised by 2 vascular strands of the stipe fusing upwards in the rachis into one X-shaped strand, the tracheids in the xylem not being intermixed with xylem parenchyma. In Athyrioid ferns, the 2 vascular strands fuse into I, U or V-shaped strands and the tracheids are intermixed with xylem parenchyma. The rhizome is dictyostelic in both. *Diplazium esculentum* (Retz.) Sw. showed presence of accessory medullary bundles in the rhizome (dictyostelic). *Diellia erecta* Brack. showed monostelic leaf trace.

Bir (1969), further, studied the cytological evolution in the Himalayan species of the genus *Diplazium*. Sixteen species were studied. Cytologically, *Diplazium* has n=41 (ct. *Athyrium* with n=40) chromosomes. About 25 species of *Diplazium* occur in the Himalayas, about a dozen species are diploid sexual (n=41). *D. polypodioides asperm* is a diploid hybrid (2n=82). The apogamous species *D. stenochlamys* C. Chr. is diploid (2n=82) and *D. simplicivenium* Holtt. is triploid with 2n=123. In *D. maximum* C. Chr. two, triploid and tetraploid, apogamous races are seen with 2n=123 and 2n=164. These taxa occur in Sikkim. The genus is regarded to be in an active state of evolution and speciation.

Singh and Roy (1969) described the spore morphology of some ferns of Pachmarhi (Madhya Pradesh). Nonperinate tetrahedral and trilete spores were seen in *Ophioglossum nudicaule*, *O. reticulatum*, *Osmunda regalis*, *Lygodium flexuosum*, *Alsophila glabra* and *Dicranopteris linearis*. Whereas *Cheilanthes farinosa* and *C. tenuifolia* showed trilete, sometimes alete, spherical and perinate spores and were distinctly larger than those of *Adiantum*, *Pteris biaurita* had perinate, trilete, bluntly triangular spores and *Sphenomeris chusana*, monolete, bilateral, non-perinous spores. Though Athyrioid fern spores are generally perinate, Pachmarhi species (*Diplazium esculentum*, *Athyrium hohenackerianum*) had non-perinate, monolete, bilateral spores. The spores of *Equisetum debile* are globose, alete, with a thin wall and a faint skin like perinous layer and measure 52×53 microns. The authors conclude that the criterion of spore morphology of Pachmarhi ferns is usable only in a limited sense in determining taxonomical affinity but can be of help in understanding the micro fossils of the place (Upper Gondwana Land). There is also no correlation between the similarity in base numbers and spore morphology, e.g. *Adiantum* and *Cheilanthes* have 29 and 30 as base numbers but very different spore morphology.

Paliwal[1] and Lalita Kakar (1967) found in the mature leaf of *Tmesipteris tannensis* the epidermal cells to be conspicuously large, with sinuous walls and thick striated cuticle. The authors noted the presence of stomata on both the adaxial and abaxial surfaces as against Sykes's (1908) observation of their occurrence on the adaxial surface only. The stomata are, however, absent along the midrib, margins, and at the tip. A substomatal chamber is generally missing. The mature stomata are of anomocytic type and follow haplocheilic mode of ontogeny as already reported by Sykes (Ann. Bot. *22*, 63, 1908).

[1]Paliwal, G. S. and Lalita Kakar, 1967. The stomata of *Tmesipteris tannensis* Bernh. Curr. Sci. 36, 412-13.

## REFERENCES


| | |
|---|---|
| Bhambi, S. and Prakasa Rao, C. G. (1969). | Root-apex organization in pteridophytes. |
| Bir, S. S. (1969). | Cytological evolution in the Himalayan species of the genus *Diplazium* Swartz. |
| Bir, S S. (1969a). | Role of anatomy in the taxonomy of asplenioid and athyrioid ferns. |
| Mehra, P. N. and Gupta, Asha (1969). | Prothalli of some Cryptogrammaceae. |
| Pal, Sunanda (1969). | Homology in the development of dermal appendages and soral structures in Polypodiaceae. |
| Panigrahi, G. (1969). | Some recent advances in the studies on Indian pteridophytes. |
| Pant, D. D. and Khare P. K. (1969). | Epidermal structure and stomatal ontogeny in *Tmesipteris tannensis* Burnh. |
| Sharma, D. N. and Tripathi, S. M. (1969). | Ontogeny of leaf buds in *Adiantum caudatum* Linn. |
| Singh, J. B. and Roy, S.K. (1969). | The spore morphology of some ferns of Pachmarhi. |
| Verma, S. C. (1969). | Genetics of gametophytic variation and mating system in *Adiantum capillusveneris* L. |
| Verma, S. C. and Goloknath Molina (1969). | Morphology and breeding system of *Cheilanthes tenuifolia* (Burm.) Sw. |

(Papers read in the Seminar on Morphology, Anatomy and Embryology of Land Plants, held on December 23-28, 1969 at the Centre of Advanced Study in Botany, Department of Botany, University of Delhi, Delhi-7).

## APPENDIX II

*Distribution of Marsilia species in India, Pakistan, Burma and Ceylon*

The information regarding the distribution of the species has been obtained from the herbaria of Kew, Calcutta, Dehra Dun, Peradeynia (Ceylon) and relevant published literature. Some intergrading species like *M. gracilenta* and *M. brachypus* cause some difference of opinion.

(Vide plates I-V.)

1. *M. minuta* L. (Syn. *M. dentata* Roxb ).

*Ostindien*[6] : Plains of India.[5] *Assam* : *Sylhet*.[1] *Bengal* : W. Duars-Moojnaie river,[1] Indian Botanic Gardens (Raizada),[3] Tipperah plain[2] (Debbarman), Old Agartala,[2] Santal Paraganas—Gauri Pahari (G. S. Gamble).[2] *Bihar* : Chhota Nagpur[1] (J. J. Wood), Singhbhum[2] (H. H. Haines, J. S. Gamble), Baloomath.[1] *Orissa* : Ganjam[1], [2], [3] Gangapur[1], [2], [3] (J.S. Gamble), Dulakundar[3] (Mooney), Chatrapur,[1] Gullery[1] Palampur[3] (G. A. Gammie). *U. P.* : Gorakhpur[1], [2] (Bhagat Ram), Jaunsar[3] (J. S. Gamble), Kheri District[3]—Bhira[3] (Inayat), Dehra Dun (Dhakaram),[3] Mussoorie[3]—Garta Hills (Duthie), Banda[3] (Mrs. Bell, 127, 831906), Bundelkhand—Konch[3] (J. F. Duthie, 66016). *Punjab* : Jullunder District (Prof. Kashyap).[2] *Rajasthan* : Ajmer Marwara[3] (J. F. Duthie), Udaipur.[3] *Delhi* (near Gurukul).[8] *S. India* : Chingleput District,[2] Coimbatore District,[2] Dharapuram,[2] Udumalpet,[2] Mysore,[2] Saidapet,[1] Godavari District[1] (J. S. Gamble), Bandamcherla[1] (near Ellore), Tanjore District[1] (Poodoo Cheltum), Negapatam,[1] Nilgiris.[1] Pondicherry.[1] *Bombay Province*[1], [9] : Khandela Pool[3] (H. Santapau, 969). *Pakistan* : Lahore[2]—Shalamar.[3], [2] *Burma*: Pegu[6] (Kurz, 3192). *Ceylon* : Dambane, Kurunjala and N. Province.[4]

2. *M. major* (Haines) Chowdhury* stat. nov. (Syn. *M. minuta* L. var. *major* Haines, Botany of B. & O., 1924, p. 1219 ; *M. quadrifoliata* non L. Prain, Bengal Plants.)

*Provisional name, subject to further studies.

*Bihar :* Chhota Nagpur—Singhbhum District[1,2] (H. H. Haines), Horhap,[1] Jashpurnagar,[1] Karbang.[1] *U. P. :* Aligarh District,[3] Meerut District,[3] Jumna Banks[3]—Agra, Delhi, Lucknow District[1] (Chowdhury). *Panjab :* Karnal.[1] *Kashmir :* Beraron and Lopour.[1] *S. India :* Kistna District,[1] Bezwada,[1] Godavari District, Tapeswaram,[1] Dumagudum.[1] *Bombay :* Poona,[1] N. Kanara,[1] Gokuru.[1]

3. *M. quadrifoliata* L.

*Bengal :* Jalpaiguri Terai[2] (Riboo and Romu), Agartala—Tipperah[2] (Debbarman). *Bihar :* Chhota Nagpur[2] (J. J. Wood). *Orissa :* Ganjam District.[2] *U. P. :* Kheri—Bhira[3] (Duthie), Banda.[2] *Kashmir.*[2] *S. India :* Mysore (Meebold)[2] *Pakistan :* Peshawar,[2] Hazara.[2] (Stewart), Kohat.[2] *N. W. F. P. Burma :* Sagain,[2] Inle Lake[2] (Anandale), S. Shan State,[2] Yomah (Kurz).

4. *M. coromandelica* Burm. (Syn. *M. coromandelina* Willd., *M. minuta* L. var. *coromandelina*).

*Ostindien*[6] *:* Plains of Pennisular India.[5] *Bengal :* Kawnia.[1] *S. India :* Near Madras,[1] Pondicherry.[1] *Bombay Province :* N. Kanara,[1] Gokuru,[1] Coomptu.[1]

5. *M. quadrifolia* L.

*Rajasthan :* Jodhpur (K. M. Gupta). *Bombay Province.*[9] Ahmedabad (Mahabale).

6. *M. brachypus* A. Br.

*Ostindien.*[6] *S. India :* Nilgiris (Wight).

7. *M. condensata* Bak.

*Bombay Province :* Scinde[1] (? Sindh, Pakistan), *Rajasthan*[10] (Nimla, Barmer).

8. *M. aegyptiaca* Willd.

*Rajasthan :* Jodhpur (K. M Gupta).[10]

9. *M. gracilenta* A. Br.

*Ostindien*[6] *Bombay Province :* Konkan.[5]

10. M. *poonensis* Kolh.

*Bombay Province :* Poona.

11. *M. ballardii* Gupta.

*Rajasthan :* Ajmer (K. M. Gupta).[10]

12. *M. brachycarpa* A. Br.
*Burma* : Pegu.[6]

---

1. Kew Herbarium.
2. Calcutta Herbarium.
3. Dehra Dun Herbarium.
4. Peradeniya Herbarium.
5. Baker (1887).
6. Engler and Prantl (1902).
7. Rabenhorst, L. (1889). Kryptogamen Flora III. Leipzig.
8. Delhi University Herbarium (Delhi Flora by J. K. Maheswari-1963-1965). C.S.I.R. publication, New Delhi.
9. Bombay Province stands for Maharashtra and Gujarat. Wordings are retained as found from Herbaria and literature.
10. '*Marsilea*' by K. M. Gupta, Botanical Monograph No. 2. 1962 C.S.I.R. Publication, New Delhi.

# BIBLIOGRAPHY*

Abeywickrama, B.A., 1956, "The genera of Ceylon pteridophytes," Ceylon J. Sci. **13** : 1-30.

Abraham, A. and Ninan, C.A., 1954. The chromosomes of *Ophioglossum reticulatum* L. Curr. Sci. **23** : 213-214.

———, 1958. Cytology of *Isoetes*. Curr. Sci. **27** : 60-61.

Abraham, A., Ninan, C. A. and Mathew, P.M., 1962. Studies in the cytology and phylogeny of the pteridophytes. J. Ind. bot. Soc. XV : 339-421.

Agharkar, S. P., 1938. Progress of botany during the past twenty-five years. In the "Progress of Science in India," edited by Prasad, B., Indian Sci. Congr. Association, Calcutta.

Ahmed, G., 1941. (a) Effect of light intensity and temperature on the growth of *Azolla filiculoides*. Proc. 28th Indian Sci. Congr. (Banaras) : 163-164.

———, 1941. (b) The growth of *Azolla filiculoides* in mineral solution without the addition of "auximone." Proc. 28th Indian Sci. Congr. (Banaras) : 164.

Ahmad, Shamsul Islam, 1952. *Alsophila glabra* (Bl.) Hook. from Sylhet. Sci. and Cult. **17** : 473.

Alston, A. H. G., 1931. "Supplement to the Handbook of Flora of Ceylon."

———, 1938. "The Kandy Flora," Ceylon.

———, 1945. An enumeration of the Indian species of *Selaginella*. Proc. nat. Inst. Sci. India, **11** : 211-235.

Baker, J. G., 1887. "Hand book of the Fern Allies." George Bell and Sons : London.

Bancroft, N., 1914. Note on the vegetative reproduction in some Indian *Selaginellas*. Ann. Bot. Lond. **28** : 680-693.

Banerji, M. L., 1951. Abnormal spike of *Helminthostachys zeylanica* L. Curr. Sci. **20** : 213.

———, 1962. An unreported fungus on *Lycopodium annotinum* L. Bull. Bot. Soc. Bengal **16** : 45-46.

*A few of the references are not indicated in the text.

Banerji, S.K. and Majumdar, G.P., 1952. The origin and development of the shoot apex in *Equisetum debile* Roxb. Proc. 39th Indian Sci. Congr. (Calcutta) : 39-40.

———, 1952. The origin and development of the shoot apex in *Lycopodium phlegmaria*. L. Proc. 39th Indian Sci. Congr. (Calcutta): 40.

Beddome, R. H., 1865-1870. "The Ferns of British India." I-XXIII Madras.

———, 1863-1873. "The Ferns of Southern India." Madras.

———, 1876. "Supplement to Ferns of Southern and British India." Madras.

———, 1883. "A Hand book to the Ferns of British India, Ceylon and the Malaya Peninsula." Thacker Spink & Co. : Calcutta.

———, 1892. "Supplement to the Ferns of British India, Ceylon and Malaya Peninsula." Thacker Spink & Co.: Calcutta.

———, 1892. "A Handbook to the Ferns of British India, Ceylon and Malaya Peninsula," with supplement. Thacker Spink & Co.: Calcutta.

Bell, P., 1953. Morphology and phylogeny of Ferns. Sci. Progr. **41** : 110-116.

Bhambi, S., 1957. Studies in Pteridophytes. I The shoot apex of *Isoetes coromandelina* L. J. Ind. bot. Soc. 36 : 491-502.

———, 1962. Studies in Pteridophytes. II Contribution to the anatomy of the axis of *Isoetes coromandelina* L. and other species. Proc. Ind. Acad. Sci. **56** : 57-75.

———, 1963. Studies in Pteridophytes. III On the structure and development of the leaf and sporophyll of *Isoetes coromandelina* L. Proc. nat. Inst. Sci India. **29B** : 169-190.

———, 1965. Studies in Pteridophytes. IV The development, structure and arrangement of leaves in some species of *Lycopodium*. Proc. Ind. Acad. Sci. **61** : 242-252.

——— and Puri, V., 1963. Shoot apex organization of Lycopodiales. *(Selaginella, Lycopodium, Isoetes.)* Memoir Ind. bot. Soc. No. 4 : 55-63.

Bharadwaja, G., 1935. The occurrence of *Isoetes* in India. Curr. Sci. 3 : 300.

Birdwood, H. M., 1887-1897. A catalogue of the flora of Mahabaleshwar and Matheran. J. Bombay nat. Hist. Soc. vol. 2, p. 107. vol. 10. p. 394. 32 species of Ferns.

Biswas, K., 1932. Glimpses of vegetation of Burma. J. Bombay nat. Hist. Soc. 36.

———, 1942. Observations on a common tree fern of the Eastern Himalayas. Proc. 29th Indian Sci. Congr. (Baroda) : 145.

———, 1943. Systematic position of a common tree fern of the Eastern Himalayas with a note on the genus *Cyathea*. J. Asiat. Soc. Beng. **9** : 43-47.

——— and Calder, C. C., 1937. "Hand book of common water and marsh plants of India and Burma." Malaria Bureau. No. **11** (Health Bulletin No. 24) : 106-109 (*Isoetes*, *Marsilia*, *Azolla*, *Salvinia* and *Ceratopteris*).

Biswas, K. P., 1935. Ferns and Fern Allies of Burma. Proc. 22nd Indian Sci. Congr. (Calcutta) : 255.

———, 1939. Two new ferns from India. Kew Bull. Misc. **27** : 237-241.

Bitter, G., 1900. Marattiaceae. In Engler, A. and Prantl, K. "Die Naturlichen Pflanzen familien." I : 444.

———, 1900. Ophioglossaceae. In Engler, Prantl's Die Naturlichen Pflanzen familien, I : 449-472.

Blatter, E. 1908. Ceylon Ferns. J. Bombay nat. Hist. Soc. **18** : 639-648.

———, 1909. The Ferns of the Bombay Presidency. J Bombay nat. Hist. Soc. **18** : 599-612.

——— and d'Almeida, J. F., 1922. "The Ferns of Bombay." D. B. Taraporevala Sons & Co. : Bombay.

Bose, S. R., 1931. Golgi bodies in *Equisetum.* Ann Bot., Lond. **45** 303-314.

Bower, F. O., 1926. The Ferns (Filicales), vol. II. Cambridge University Press : London.

———, 1935. Primitive Land Plants. McMillan & Co. : London.

Burkill, I. H., 1925. The botany of Abor (Selaginellaceae). Rec. bot. Surv. India **10** : 227-228.

———, 1925. The Botany of Abor expedition. Rec. bot. Surv. India, **10**.

Chakravarty, H. L., 1951. Indian *Ophioglossum.* Bull. Bot. Soc. Beng. **5** : 1-10.

Chamberlain, C. J., 1935. Methods in plant histology. Chicago.

Chatterjee, Jyotsna and Mohan Ram, H. Y., 1968. : Gametophytes of *Equisetum ramosissimum* Desf. I Structure and Development. Bot. Notiser **121** : 471 – 490. Sweden.

Cheema, G. S., 1918. The leaf trace in Polypodiaceae. Proc. 5th Indian Sci. Congr. (Lahore) : 97.

———, 1919. The rhizome and frond of *Goniopteris prolifera* Roxb. Proc. 6th Indian Sci. Congr. (Bombay) : 169.

Chopra, R. N., 1933. "Indigenous Drugs of India." The Art Press: Calcutta.

Chowdhury, N., 1937. Notes on Indian species of *Lycopodium* with remarks on the distribution of the genus in India, Burma, and Ceylon. Trans. nat. Inst. Sci. India **1** : 187-226.

———, 1943. Studies on Indian xerophytic species of *Selaginella.* Proc. 30th Indian Sci. Congr. (Calcutta) : 53.

Chowdhury, N. P., 1932. On the occurrence of superficial sori in *Osmunda claytoniana* L. Proc. 19th Indian Sci. Congr. (Bangalore) : 307-308.

———, 1932. On the occurrence of superficial abaxially placed sori in *Osmunda claytoniana* L. J. Indian bot. Soc. **11** : 137-145. (Also American Fern J. **22** : 132).

———, 1933. A study of the genus *Lycopodium* in India. Proc. 20th Indian Sci. Congr. (Patna) :

———, 1957. Observations on the structure and ecology of a xerophytic *Selaginella* from India and distribution pattern of xerophytic *Selaginella* in India, Burma and Ceylon (*S. Wightii* Hieron.). Bull. Bot. Soc. Bengal **11** : 41-48.

———, 1959. Observations on the structure and ecology of a xerophytic *Selaginella* from India *I. S. bryopteris* (L.) Baker. Jap. Jour. Bot. Tokyo **17** : 101-119.

Chowdhury, N. P., 1971. Researches on living Pteridophytes in India, Burma and Ceylon. Asia Publishing House, Bombay.

——— and Raizada, M. B., 1954. Ferns and Fern Allies of the Upper Gangetic Plain. Proc. 41st Indian Sci. Congr. (Hyderabad) : 118.

———, 1961. The Fern Allies of the Upper Gangetic Plain. Bull. Bot. Soc., Univ. of Saugor. **13** : 31-43. (Prof. S. K. Pande Commemoration Volume.)

——— and Mehra, P. N., 1957. Progress of Science in India. VI. Botany (Pteridology). Nat. Inst. Sci. India Publication, Calcutta.

Clarke, C. B., 1880. A review of the Ferns of Northern India. Trans. Linn. Soc. Lond. (Bot.) **1** : 425-611.

Dalgado, D. G., 1896. Note on *Psilotum triquetrum*. J. Bombay nat. Hist. Soc. **7**: 544.

D' Almeida, J. F. R., 1922. The Indian *Ophioglossums*. J. Indian bot. Soc **3** : 58-65.

———, 1926. A new species of *Nephrolepis*. J. Indian bot. Soc. **5** : 51-54.

Dalzell, N. A., 1852. Contributions to the botany of Western India. J. Bot. Lond. **4** : 107-114.

Deshpande, J. V., 1943. A study of the sporophyte of *Salvinia cucullata* Roxb. J. Indian bot. Soc. **22** : 59-84.

Dickason, F. G., 1946 The ferns of Burma. Ohio. J. Sci. **46** : 109-141.

———, 1946. A phylogenetic study of the Ferns of Burma. Ohio J. Sci. **46** : 73.

———, 1946. A phylogenetic study of the Ferns of Burma. Abst. Doct. Dissertations No. 50, Ohio State University Press, pp. 1—16.

Diels, L., 1900. Parkeriaceae. *In* Engler, A. and Prantl, K. "Die Naturlichen Pflanzenfamilien." **1** : 339-342.

———, 1902. Cyatheaceae. *In* Engler, A. and Prantl, K. "Die Naturlichen Pflanzenfamilien." **1** (4).

———, 1902. Polypodiaceae *In* Engler, A. and Prantl, K. "Die Naturlichen Pflanzenfamilien." **1** : 139-339.

Dixit, D. L., 1934. Polarity of sexual reproduction in *Selaginella*. Proc. 21st Indian Sci. Congr. (Bombay) : 300.

——— and Mahabale T. S., 1934. Germination of the spores of *Ophioglossum* of Poona. Proc. 21st Indian Sci. Congr. (Bombay) : 299.

———, 1934. *Ophioglossum* species near about Poona. Proc. 21st Indian Sci. Congr. (Bombay) : 300.

Dutt, B. S. M., 1950. Occurrence of *Psilotum nudum* L. Griseb. (*Psilotum triquetrum* Sw.). Curr. Sci. **19** : 227.

Dymock, W., Warden, J. H. and Hooper, D., 1890-1893. "Pharmacographia Indica." Thacker Spink & Co. : Calcutta.

Eames, A. J., 1936. Morphology of vascular plants (lower group), New York.

Ekambaram, T. and Venkatanathan, T. N., 1933. Studies on *Isoetes coromandelina* L. J. Indian bot. Soc. **12** : 191-225.

Engler, A. and Prantl, K., 1900-1902. "Die Naturlichen Pflanzen-familien." I. Verlag von Wilhelm Engelmann : Leipzig.

Fergusson, W., 1886. The vascular cryptogams of Ceylon. Colombo.

Gaekwad, L. K., 1954. Occurrence of *Isoetes* at Baroda in Gujarat Prov. Bombay State. Sci. and Cult. **19** :

Ghosh, A. K., 1941-42. On the occurrence of two sporangiate sporophylls in *Lycopodium phlegmaria* L. Sci. and Cult. **7** : 410-411.

Ghosh, S. L., 1916, 1932. *Selaginella* from Garhwal. J. Bombay nat. Hist. Soc. **24** : 16. Proc. 19th Indian Sci. Congr. (Bangalore) : 308.

———, 1920. Evolution in the cone of *Selaginella*. Proc. Lahore Phil. Soc. 2 : 1-9.

Goebel, K., 1905. Organography of Plants. II. Clarendon Press, Oxford (English edition).

Gopal, B., 1966. Vegetative propagation in *Marsilia minuta L.* Journ. Ind. bot. Soc. **45** : 350-352.

Govindu, H. C. and Narayanan, H. S., 1949. Phyllody in *Botrychium virginianum* Sw. Sci. and Cult. **15** : 199-200.

Gray, W., 1886. The botany of the Bombay Presidency. Bombay Gazetteer **25** : 376-77. (Mentioned 50 species of ferns).

Griffith, 1908. Rec. bot. Surv. India **3** : 327.

Haines, H. H., 1924. Botany of Bihar and Orissa (Cyatheaceae). **6**: 1183.

———, 1924. "The botany of Bihar and Orissa" (Pteridophytes). **6** : Adlaid & Son and West Newman Ltd. : London.

Handa, K. L., Kapoor, L. D. and Chopra, I. C., 1947. Male Ferns of Kashmir. Curr. Sci. **16** : 56-57.

Hans Raj, 1923. Observations on *Lygodium*. Proc. Lahore phil. Soc. **4** : 29-48.

Hieronymus, G., 1900-1901. Selaginellaceae. In Engler, A. and Prantl, K. "Die Naturlichen Pflanzenfamilien." **1** : 625-720.

Holloway, G.E., 1915, Heterophyllous species of *Lycopodium*. Trans. New Zealand Insti. 48:

———, 1917. Seedlings of *Tmesipteris.* Trans. N. Z. Insti. 50 : 1-44.

Hope, C. W., 1890. Three new *Lastreas* from Assam. J. of Bombay, 1890 : 1-3.

———, 1899. The Ferns of North-Western India. J. Bombay nat. Hist. Soc. **12** : 315-527.

———, 1903. The ferns of Northern India. J. Bombay nat. Hist. Soc. **15** :

——— and Trotter, E. W., 1890. "Catalogue of Ferns in the Herbarium of Govt. of India at Saharanpur." (Species of Trans. Indus region, India and Burma).

Javalgekar, S. R., 1960. Sporogenesis and prothallial development in *Ceratopteris thallictroides.* Bot. Gaz. **122** : 45-50.

——— and Mahabale, T. S., 1959. Germination of spores and prathalli in two species of *Nephrolepis*, *N. exaltata* Schott. and *N. acuta* Presl. Proc. nat. Insti. Sci. India. **25** (B) : 333-338.

Johnson, D. S., 1898. On the leaf and sporocarp of *Marsilia quadrifolia.* Ann. Bot. London. **12** : 119-145.

Joshi, P. C., 1935. Occurrence of *Psilotum* Sw. in the Punjab. Curr. Sci. **3** : 486.

Kachroo, P., 1952. Spore germination in *Alsophila glabra* Hk. Sci. and Cult. **17** : 342.

———, 1953. Ferns of Assam. J. Asiat. Soc. Beng. **19** : 161-174.

Kachroo, P., and Nayar, B. K., 1952. Morphology of *Adiantum* (*A. aethiopicum*). J. Univ. Gauhati **3** : 163-171.

———, 1953. Studies in Pteridaceae. Observations on gametophytes of some species of *Adiantum* L. Phytomorph. **3** : 240-248.

Kapil, R.N., 1968. Some contributions of Professor P. Maheshwari to Botany. Phytomorphology **17** : 1-23.

Kashyap, S. R., 1914. The structure and development of the prothallus of *Equisetum debile* Roxb. Ann. Bot. Lond. **28** : 163-181.

———, 1917. Notes on *Equisetum debile* Roxb. Ann. Bot. Lond. **31** : 439-445.

———, 1930. Some abnormal cones in *Equisetum debile.* J. Indian bot. Soc. **9**.

———, 1932. The gametophyte of *Equisetum debile* Roxb. J. Indian bot. Soc. **11** : 86-88.

——— and Mehra, P. N., 1933. Dichotomous branching in the leaves of *Pleopeltis simplex* Sw. Proc. 20th Indian Sci. Congr. (Patna): 309.

———, 1934. Dichotomous branching in the leaves of *Pleopeltis simplex* Sw. Curr. Sci. 3: 207.

Kingdonward, F., 1944-45. A sketch of the botany and geography of N. Burma. Pts. I-III. J. Bombay nat. Hist. Soc. **44** : **45**.

Kirtikar, K.K. and Basu, B. D., 1941. "Medicinal Plants of India." Vol. IV : 2nd edition. Indian Press : Allahabad.

Kolhatkar, G. G., 1925. The study of the female gametophyte and megasporogenesis of *Marsilia* from Poona. Proc. 12th Indian Sci. Congr. (Banaras) : 301.

———, 1934. Sporogenesis of *Marsilia* of Poona. Proc. 21st. Indian Sci. Congr. (Bombay) : 301

———, 1937. The study of the male gametophyte and spermatogenesis of *Marsilia* from Poona. Proc. 24th Indian Sci. Congr. (Hyderabad) : 264.

———, 1937. On life history of *Marsilia* from Poona. J. Univ. Bombay. 5 : 19-37.

———, 1940. A new species of *Marsilia* from Poona, *Marsilia poonensis*. Proc. 27th. Indian Sci. Congr. (Madras) : 301.

Kurz, S., 1877. "Forest Flora of British Burma." 1-11.

Lewis, F., 1934. "Vegetable Products of Ceylon." 21st Indian Sci. Congr. (Bombay).

MacCann, C., 1934. Occurrence of *Isoetes coromandelina* L. J. Bombay nat. Hist. Soc. **37** : 501-502.

MacMillan, H. F., 1943. "Tropical Planting and Gardening." McMillan & Co. : London.

Macpherson, T. R. M., 1894. List of Ferns gathered in N. Kanara. J. Bombay nat. Hist. Soc. **5** : 375.

Mahabale, T. S., 1937. Dichotomy as the probable cause of the so-called abnormalities in *Ophioglossum*. Proc. 24th Indian Sci. Congr. (Hyderabad) : 264.

———, 1937. The fertile spike of *Ophioglossum*. Proc. 24th Indian Sci. Congr. (Hyderabad) : 264-265.

———, 1937. The gametophyte of *Ophioglossum aitchisoni* d' Alm. Proc. 24th Indian Sci. Congr. (Hyderabad) ; 265.

———, 1937. The gametophyte of *Ophioglossum fibrosum* Schum. Proc. 24th Indian Sci. Congr. (Hyderabad) : 265.

———, 1937. The gametophyte of *Ophioglossum pendunculosum* Desv. Proc. 24th Indian Sci. Congr. (Hyderabad) : 265-266.

———, 1937. On the discovery of the prothallus of *Lycopodium* in India. J. Indian bot. Soc **16** : 145-149.

———, 1938. Distribution of the Psilotaceae, the Equisetaceae, and the Lycopodiaceae with notes on the distribution of the species. J. Univ. Bombay **6** : 63-74.

———, 1938. Studies on the Vascular cryptogams of the Bombay Presidency. Pt. II. Distribution of the Ophioglossaceae with notes on the ecology of the species. J. Univ. Bombay. **6** : 104-117.

———, 1938. On a new species of *Isoetes* in India. Curr. Sci. **7** : 62-63.

———, 1945. The concept of vestigial organs and the vascular cryptogams. Proc. 32nd Indian Sci. Congr. (Nagpur) : 71-72. Also Curr. Sci. **15** : 220-225.

———, 1945. Prothallus of *Equisetum debile* var. *Pashan Bhide*. J. Univ. Bombay. **14** : 42-48.

———, 1947. Prothalli of *Gymnopteris variabilis* Bedd. var. *lanceolata*. Proc. 34th Indian Sci. Congr. (Delhi) : 152.

———, 1948. Prothalli of *Ceratopteris thallictroides*. Bot. Gaz. **109** : 349-354.

———, 1948. Prothallus of *Lycopodium hamiltonii* Spr. Curr. Sci. **17** : 187-189.

———, 1949. The laticiferous system of *Regnellidium*. Proc. 36th Indian Sci. Congr. 126.

———, 1949. The laticiferous system of *Regnellidium diphyllum* Lind. Curr. Sci. **18** : 449-50.

———, 1953. Anatomy and prothallia of *Gymnopteris quercifolia* Benth. Proc. 40th Indian Sci. Congr. (Lucknow) : 85.

———, 1962. Buds in some Indian ferns. Proc. Summer school of Botany, 1960. Darjeeling : 136-139.

———, 1963. Evolutionary tendencies in the genus *Azolla*. Memoirs Ind. bot. Soc. No. 4. (Phylogeny and evolution in pteridophytes.)

——— and Biswas, T., 1949. The structure and life history of *Anisogonium esculentum* Presl. Proc. 36th Indian Sci. Congr. 124.

——— and Chenna, M. S., 1954. Chromosomes in *Regnelltdium* and *Marsilia*. Proc. 41st. Indian Sci. Congr. (Hyderabad) : 118.

———, and d' Almeida, N., 1953. Studies on two species of *Davallia*. Proc. 40th Indian Sci. Congr. (Lucknow) : 85.

——— and Deshpande, G. S., 1934. Variability of parts in the genus *Ophioglossum*. Proc. 21st Indian Sci. Congr. (Bombay) : 300.

———, 1942. *Psilotum triquetrum* Sw. at Lonavla, Bombay Presidency. Curr. Sci. **12** : 464.

——— and d' Mello, Joyce, 1950. Massulae in the genus *Salvinia*. Proc. 37th Indian Sci. Congr. : 8.

———, 1952. Apogamy in *Salvinia*. Curr. Sci. **2** : 227-228.

——— and Gorji, G. H., 1948. Some observations on the sporelings and adult plants of *Marsilia quadrifolia* Linn. J. Univ. Bombay **16** : 27-34. (Also Biol. Abst. 1950, **24** : 691. Abs. No. 7050).

——— and Javalgekar, S. R., 1950. Germination of spores and adult prothallus in culture of *Ceratopteris thallictroides* Brogn. Proc. 37th Indian Sci. Congr. : **42.**

———, 1951. Sporogenesis in *Nephrolepis exaltata*. J. Univ. Bombay **20** ; 66-77.

——— and Javalgekar, S. R , 1959. On a tuberiferous species of *Nephrolepis*, *N. paucifrondosa* D'Almeida. J. Univ. Bombay 27 : 43-56.

——— and Kulkarni, N., 1949. Prothallus of *Lygodium flexuosum* Bedd. Proc. 36th Indian Sci. Congr. : 124.

——— and Kulkarni, H. Y., 1950. Prothalli of *Lygodium japonicum* Sw. and *Aneimia rotundifolia* Schrad. Proc. 37th Indian Sci. Congr. : 41.

——— and Shah, J. J., 1949. Prothallus of *Cheilanthes farinosa*, Kaulf and *C. tenuifolia* Sw. Proc. 36th Indian Sci. Congr. 125.

———, 1950. Contribution to the study of the genus *Cheilanthes* Proc. 37th Indian Sci. Congr. 41.

——— and Venkateswaran, S., 1953. Life History of *Athyrium hohencharianum* Bedd. Proc. 40th Indian Sci. Congr. (Lucknow) : 85.

——— and Yenkateswaran, S., 1959. Studies on the Polypodiaceae of Bombay : Morphology of *Pleocnemia membranifolia* Presl. J. Univ. Bombay 27 : 30-42.

——— and Javalgekar, S. R., 1949. Prothallus in the genus. *Nephrolepis*. Proc. 36th Indian Sci. Congr. 126.

Maheshwari, P., 1952. Prof. B. Sahni's contributions on living plants. Palaeobiotanist. **1** : 17-21.

——— and Bhargava, H. R., 1935. A solenostelic *Lindsaya*. Proc. 22nd Indian Sci. Congr. (Calcutta) :

——— and Puri, V., 1935. A note on the occurrence of a smut on the Indian Selaginellas. Curr. Sci. **3**.

——— and Singh, B. 1934. The morphology of *Ophioglossum fibrosum* Schum. J. Indian bot. Soc. **13** : 103-123.

———, 1934 The morphology of *Ophioglossum fibrossum* Schum. Proc 21st Indian Sci. Congr. (Bombay) : 299.

Majumdar, G. P., 1926. Report on the occurrence of *Equisetum debile* Roxb. on the sandy alluvial bank of a river in Pabna, lower Bengal. Proc. 13th Indian Sci. Congr. (Bombay) : 223.

———, 1930. Report on the occurrence of *Helminthostachys zeylanica* Hk. and *Equisetum debile* Roxb. in the suburbs of Calcutta. Proc. 17th Indian Sci. Congr. (Allahabad).

———, 1931. On the origin of medullation in *Selaginella* Spr., and the Stelar theory. Proc. 18th Indian Sci. Congr. (Nagpur) : 268.

———, 1933. Origin of leafy sporophyte in Pteridophytes (a discussion). Proc. 20th Indian Sci. Congr. (Patna) : **309**.

———, 1933. Role of anatomy in Plant taxonomy. Sci. and Cult. **16** : 541-542.

———, 1933. A preliminary report on the fern flora of Calcutta and its suburbs with a note (morphological and anatomical) on the vegetative propagation of *Polypodium proliferum* Roxb. Proc. 20th Indian Sci. Congr. (Patna) : 308-309.

———, 1937. Morphological nature of the compound leaves of ferns. Proc. 24th Indian Sci. Congr. (Hyderabad) : 266.

———, 1942. On the origin of Siphonostele in three species of *Selaginella* Spr. Proc. Indian Acad. Sci. : **15** : 172-76.

———, 1944. Preliminary report on the origin of the shoot borne roots on adult internodes of *Marsilia quadrifida*. Proc. 31st Indian Sci. Congr. (Delhi) : 81.

Manton, I. and Sledge, W. A., 1954. Observations on the cytology and taxonomy of the pteridophyte flora of Ceylon. Phil. Trans. Roy. Soc. Lond. 238 (Ser. B) : 127-185.

Mehra, P. N., 1931. A study of the gametophyte of *Adiantum capillus veneris* and its bearing on the systematic position of the genus.

Proc. 18th Indian Sci. Congr. (Nagpur) : 265-266. Also Curr. Sci. **1** : (1932).

———, 1931. Filamentous fern prothalli under cultural condition. Proc. 18th Indian Sci. Congr. (Nagpur) : 266. Also Curr. Sci. **1** : (1932).

———, 1931. A morphological and structural study of *Nephrolepis ramosa* Beauv. with special reference to the origin of soral receptacle. Proc. 18th Indian Sci. Congr. (Nagpur) : 266-267.

———, 1931. Coenosorii in *Nephrodium molle* Desv. Proc. 18th Indian Sci. Congr. (Nagpur) ; 267.

———, 1932. A note on the occurrence of superficial sori in *Osmunda claytoniana* L. J. Indian bot. Soc. **11** : 146-147.

———, 1932. Filamentous fern prothalli under cultural conditions. Curr. Sci. **1**.

———, 1932. *Ceropteris calomelanos* in Sikkim. J. Indian bot. Soc. **11**: 340-341.

———, 1932. Occurrence of tracheids in the gametophyte of *Adiantum lunulatum* Burm. Curr. Sci. **1**.

———, 1932. Apogamy in some Indian ferns. Proc. 19th Indian Sci. Congr. (Bangalore); 306-307.

———, 1932. Repeated dichotomy in the gametophyte of *Anisogonium esculentum*. Proc. 19th Indian Sci. Congr. (Bangalore) : 307.

———, 1932. Collenchymatous thickening in some prothallia. Proc. 19th Indian Sci. Congr. (Bangalore) : 307.

———, 1937. Cytological investigation in the apogamous life cycle of *Adiantum lunulatum* Burm. Proc. 24th Indian Sci. Congr. (Hyderabad) : 266.

———, 1938. Apogamy in *Adiantum lunulatum* Burm. Proc. Indian Acad. Sci. B. **8** : 192-200. Pt. 1, Morphology.

———, 1938. Apogamy in *Pteris biaurita* L. Proc. Indian Acad, Sci. B. **8** : 202-210.

———, 1938. Abnormal sporocarps in *Marsilia minuta* L. Proc. Indian Acad. Sci. B. **8** : 7-10.

———, 1939. "Ferns of Mussoorie." Univ. of Punjab, Lahore, Publication No. 29.

———, 1944. Rhizome anatomy of the fern *Campteria wallichiana* Proc. Lahore phil. Soc. **6** : 50-53.

———, 1945. Cytological investigation of apogamy in *Adiantum lunulatum* Burm. Proc. 32nd Indian Sci. Congr. (Nagpur) : 75-76.

———, 1949. Some observations on the Prothallus of *Diplazium esculentum* (Retz) Sw. J. Indian bot. Soc. Feb. 1949, 1-8.

———, 1949 a. Cytological investigation of apogamy in *Adiantum lunulatum*. Proc. Nat. Acad. Sci. India. **14**. 189-204.

———, 1952. Colchicine effect and the production of abnormal spermatozoids in the prothalli of *Dryopteris subpubescens* (Bl.) C. Chr. and *Goniopteris prolifera* Roxb. Ann. Bot. Lond. N.S. **16** :

Mehra, P. N., 1961. Cytological Evolution of Ferns with particular reference to Himalayan ferns. Presidential Address. Sec. Botany. 48th Indian Sci. Congr. (Roorkee) : 1-24.

———, 1968. Palynology of Archegoniatae (an evolutionary approach). Journ. Palynology. Lucknow. 4 : 56-72.

——— and Bir, S. S., 1964. Pteridophytic flora of Darjeeling and Sikkim Himalayas. Res. Bulletin (N.S.) Punjab University. **15**: 169-181.

——— and Chopra, N., 1951. The Anatomy of *Staenochlaena palustris* (Burm.) Bedd. Ann. Bot., Lond. N. S. **15** : 37-45.

——— and Chowdhury, N. P., 1957. Progress of Science in India. Sec. VI. Botany. Nat. Insti. Sci. India Publication (Pteridology).

——— and Gurdip Singh, 1957. Cytology of Hymenophyllaceae. J. Genet. 55 : 379-393.

——— and Loyal, D. S., 1959. Cytological studies in *Marsilia* with particular reference to *M. minuta* L. Punjab Univ. Res. Bull. (N.S.) 10 : 357-374.

——— and Mittal, T. C., 1961. Significance of internal secretory glands in relation to "Filicin." Planta Medica **9**: 189-199. (Stuttgart).

——— and Varma, S. C., 1957. Cytology of *Lycopodium*. Curr. Sci. 25 : 55-56.

Mehta, A. S., 1952. *Ophioglossum vulgatum*, L. Curr. Sci. **21** : 26.

Milde, J., 1867. Monographia *Equisetorium*. Nova Acta **32** : 32.

Misra, R. D., 1935. On the Benares *Isoetes*. Proc. 22nd Indian Sci. Congr. (Calcutta) : 254.

Mitra, A. K., 1944. *Melanopsamma ranjanii* Sp. nov. A new parasite of *Selaginella* J. Indian bot. Soc. **23** : 158-163.

Mitra, M., 1919. Morphology of some species of *Pteris* and *Adiantum*. Proc. 6th Indian Sci. Congr. (Bombay) : 168-169.

Mooney, H. P., 1950. Supplement to "The botany of Bihar and Orissa." Haines, H. H. (1924). Catholic Press : Ranchi.

Mulay, B. N., 1937. Chromatin material found in the microsporocarps of *Azolla pinnata*. Proc. 24th Indian Sci. Congr. (Hyderabad) : 266.

———, 1938. Development of the female prothallium in *Azolla pinnata* R. Br. J. Univ. Bombay **6** : 118-124.

———, 1939 a. Some observation on the microsporocarps, microsporangia, and the male prothallium of *Azolla pinnata* R. Br. Proc. 26th Indian Sci. Congr. (Lahore) : 119-120.

———, 1939 b. Fertilization in *Azolla pinnata* R. Br. Proc. 26th Indian Sci. Congr. (Lahore) : 120.

———, 1939 c. Some ecological observations on the life history of *Azolla pinnata* R. Br. Proc. 26th Indian Sci. Congr. (Lahore) : 126.

Nair, S. L. and Chopra, I. C., 1951. "Distribution of British Pharmacopoeal drug plants and their substitutes growing in India." C.S.I.R. Publication, New Delhi.

Nayar, B. K., 1964. Palynology of modern pteridophytes. (Recent advances in palynology). Nat. Bot. Gardn. Publication, Lucknow.

Nayar, B.K. and Kachroo, P., 1953. Studies in Polypodiaceae 1. Contribution to the morphology of *Drynaria* Bory. (*D. quericifolia* L. Sw. and *D. propinqua* L.). Phytomorphology **3** : 411.

Ninan, C. A., 1955. Cytology of *Equisetum debile* Roxb. J. Ind. Bot. Soc. 34 : 112-114.

———, 1956. Cytology of *Psilotum nudum* (L.) Beauv. (*P. triquetrum* Sw.). La Cellule 57 : 307-318.

———, 1958. Studies on the cytology and phylogeny of pteridophytes. Observations on the genus *Lycopodium*. Proc. nat. Insti. Sci. India. **24B** : 54-56.

Ninan, C. A., 1958. Studies on the cytology and phylogeny of pteridophytes. V. Observations on the Isoetaceae. Journ. Indian bot. Soc. **31** ; 95-103.

Palaeobotanist (Birbal Sahni Memorial Volume)—I ; 1952 ; 1-533 (Sahni Institute of Palaeobotany, Lucknow).

Pande, S. K., 1935. Notes on the anatomy of a xerophytic fern *Niphobolus adnascens* from the Malay Peninsula. Proc. Indian Acad. Sci. B. **1** : 556-564.

Pande, S. S., 1923. Some observations on the biology of *Marsilia*. Proc. Lahore phil. Soc. **4** : 1-28.

———, 1925. Some observations on the biology and life history of *Marsilia erosa*. Proc. 12th Indian Sci. Congr. (Banaras) : 194.

Pandeya, S. C., 1953. Variations of leaf form in *Marsilia quadrifoliata* L. Proc. 40th Indian Sci. Congr. (Lucknow) : 85.

Pandit, B. R. and Mulay, B. N., 1931. *Azolla* at Khandala. Proc. 18th Indian Sci. Congr. (Nagpur) : 267.

Panigrahi, G. and Chowdhury, S., 1962. Enumeration and distribution of fern-allies in Eastern India. Proc 44th. Indian Sci. Congr : 255.

——— and Dixit, R. D., 1966. Studies in the systematics of Indian *Selaginella*. Proc. Nat. Acad. Sci. India (B) **36** : 102-108.

——— and Manton, I., 1958. Cytological and taxonomic observations on some members of the *Cyclosorus parasiticus* complex. Journ. Linn. Soc. London **55** : 729-743.

——— and Patnaik, S. N., 1968. Studies in the family Polypodiaceae in Eastern India. Distribution and Ecology. J. Ind. bot. Soc. **47** : 1-6.

Pant, D. D. and Mehra, B., 1963. Development of stomata in *Psilotum nudum* (L.) Beauv. Curr. Sci. **32** : 420-422.

Phatak, V. G., 1931. Some new facts regarding the gametophyte of *Equisetum*. Proc. 18th Indian Sci. Congr. (Nagpur) : 258.

———, 1932. Prothallus formation in a new species of *Equisetum* growing at Pashan, Poona. Proc. 19th Indian Sci. Congr. (Bangalore) : 309.

———, 1932. A note on the sex differentiation in gametophyte of *Equisetum*. Proc. 19th Indian Sci. Congr. (Bangalore) : 308.

———, 1932. A note on the meristematic regions in the gametophyte of *Equisetum*. Proc. 19th Indian Sci. Congr. (Bangalore) : 309.

———, 1935. Researches sur une especee d' *Equisetum* recontree dans la Presidency de Bombay (Dekkan). A thesis presented to the Faculty of Science, University of Montpellier, Montpellier, 1935 : 1-70.

Prain, D., 1903. 'Bengal Plants.' **11** : 1237-1275.

———, 1908. The vegetation of the districts of Hughli, Howrah and 24 Parganas. Rec. bot. Surv. India. **3**.

Pritzell, E., 1902. Pteridophyte *(Lycopodium)*. In Engler, A. and Prantl, K. "Die Naturlichen Pflanzenfamilien." **1** : 563-606. Verlag von Wilhelm Engelmann, Leipzig.

Puri, V. and Garg, M. L., 1953. A contribution to the anatomy of the sporocarp of *Marsilia minuta* L. with a discussion on the nature of sporocarp in the Marsiliaceae. Phytomorphology **3** : 190-209.

Raizada, M. B., 1935. The genus *Psilotum* in India. Indian For. **61** : 654-657.

———, 1952. Ferns of Nepal. Indian For. **78** : 576-581.

——— and Chowdhury, N. P., 1954. Ferns and Fern Allies of the upper Gangetic plain. (i) *Psilotum*, (ii) Selaginellaceae, (iii) Isoetaceae, (iv) Equisetaceae, (v) Ophioglossaceae, (vi) Schizaeaceae, (vii) Marsiliaceae, (viii) Salviniaceae. Proc. 41st Indian Sci. Congr. (Hyderabad) : 118.

——— and Chowdhury, N. P., 1961. The Fern Allies of the upper Gangetic plain (Psilotaceae, Selaginellaceae, Isoetaceae and Equisetaceae). Bull. bot. Soc. Univ. of Saugor, **13**: 31-43. (Prof. S.K. Pande commemoration volume).

Raghavan, V., 1964. Differentiation in Fern Gametophytes treated with purine and pyrinidine analogs. Science 140 : 1690-91.

Rao, A. R., 1946. Sporophyte and anatomy of *Hemionitis arifolia* Sw. Curr. Sci. **15** : 141-142.

———, 1949. The prothallus of *Hemionitis arifolia* Sw. Curr. Sci. **18** : 349-350.

Rao, H. S., 1935. The structure and life history of *Azolla pinnata* R. Br. with remarks on the fossil history of Hydropterideae. Proc. Indian Acad. Sci. B. **2** : 175-200.

Rao, L. N., 1939. A note on the gametophyte of *B. virginianum* var. *lanuginosum* Sw. Curr. Sci. **8** : 119.

———, 1944. New species of *Isoetes* from Bangalore, Mysore State. Curr. Sci. **13** : 286-287.

———, 1960. Some abnormalities in *Helminthostachys zeylanica*. Curr. Sci. **29** : 190-191.

———, 1961. The archegonial complement of *Botrychium lanuginosum* Wall. ex Hook. et Grev. Curr. Sci. **30** : 388-389.

Roxburgh, W., 1832. "Flora Indica." **3** : 745.

Roy, S. K., 1964. A chromosome count in *Mohria*. Nature **201** : 348.

——— and Holttum, R. E., 1965. New cytological records for *Cystodium* and *Dicksonia*. Amer. Fern. Journal **55** : 35-37.

Saha, B., 1963. Morphogenetic studies on distribution and activities of leaf meristems in ferns. Ann. Bot. **27** : 269-78.

Sahni, B., 1915. The Anatomy of *Nephrolepis volubilis* J. Sim. with remarks on the biology and morphology of the genus. New Phytol. **14**: 251-274.

———, 1916. The vascular anatomy of the tubers of *Nephrolepis*. New Phytol. **15** : 72-80.

———, 1917. Observations on the evolution of branching in the Filicales. New Phytol. **16** : 1-23.

———, 1921. On a new abnormality in the sporophyll of *Tmesipteris*. Proc. 8th Indian Sci. Congr. (Calcutta). Also J. Asiat. Soc. Bengal N. S. **17** : 179.

———, 1923. Modern Psilotaceae and archaic terrestrial plants. Nature. Lond. **111** : 84.

———, 1923. On the theoretical significance of certain so-called "*Abnormalities*" in the sporangiophores of the Psilotaceae. J. Indian bot. Soc. 3 : 185-191.

———, 1925. On *Tmesipteris vieillardii* Dangeard. Phil. Trans. Roy. Soc. London. **213 B** : 143-170.

———, 1925. The ontogeny of vascular plants and the theory of recapitulation. J. Indian bot. Soc. **4** : 202-216.

———, 1925. Presidential Address. Sec. Bot. 12th, Indian Sci. Congr. (Calcutta).

———, 1936. The occurrence of *Matonidium* and *Weichselis* in India. Rec. geol. Surv. India **71** ; 152-165.

———, 1946. A museum of evolution. Curr. Sci. **15** : 99-100.

Sahni, B. 1948. The prospects of palynology in India. Svensk. Bot. Tidskrift **42** : 474-477.

——— and Pande, S. K., 1924. Notes on the anatomy of a xerophytic fern *Niphobolus adnascens* from the Malay Peninsula. Proc. 11th Indian Sci. Congr. (Bangalore) : 141.

Saini, H. R., 1924. The anatomy of the sporophyte and the development of the sporangium in *Lygodium japonicum* Sw. Proc. 11th Indian Sci. Congr. (Bangalore) : 140-141.

Saksena, R. K. and Mathur, L. N., 1925. On an Indian *Ophioglossum*. J. Indian bot. Soc. **4** ; 307-311.

Sampath Kumaran, M.A. and Gopal Iyenger, A.R., 1932. Life History of *Gleichenia dichotoma.* Pt. 1. Proc. 19th Indian Sci. Congr. (Bangalore): 334-335.

Sarbadhikari, P. C., 1924. Cytology of *Osmunda* and *Doodia*. (I. On the somatic and meiotic mitoses of *Doodia*). Ann. Bot. Lond 38 : 1-26.

———, 1927. Cytology of *Osmunda* and *Doodia* (II. On the gametophyte and post meiotic mitoses in the gametophytic tissue of *Doodia*). Ann Bot. Lond. 41: 1-35.

———, 1935. On the chromosome constitution of the meiotic chromosomes in *Osmunda javanica* Bl. Proc. 22nd Indian Sci. Congr. (Madras) : 255.

———, 1936. Significance of Chromatic bodies in *Osmunda javanica* Bl. Proc. 23rd Indian Sci. Congr. (Lucknow): 293.

———, 1937. Vascular system of *Osmunda cinnamonea.* Proc. 24th Indian Sci. Congr. (Hyderabad): 266.

———, 1949. Cytology of apospory in *Doodia aspera.* Proc. 36th Indian Sci. Congr. 138.

———, 1950. Cytology of apospory in *Doodia aspera.* Proc. 37th Indian Sci. Congr. 43.

Sen, B., 1937. The effect of different cations on the living protoplasm of the root hair of *Azolla pinnata* R. Br. Indian J. agric. Sci. **7** : 479-485.

Sen, S. and Nandi, P., 1951. Antibiotics from the pteridophytes. Sci. and Cult. **16** : 328-329.

Sethi, M. L., 1926. The archegonium, fertilization and after in *Equisetum debile.* Proc. 13th Indian Sci. Congr. (Bombay): 219-220.

———, 1928. Contributions to the life history of *Equisetum debile* Roxb. Ann. Bot., Lond. **42** : 729-738.

Sharma, U., 1958, 1959. Contributions to our knowledge of *Isoetes sampatkumarani*. I. (vegetative parts). Proc. Ind. Acad. Sci. **47** : 210-214. II. (Megasporogenesis) Proc. Ind. Acad. Sci. **50** : 319-339.

Sharma, U., 1960. Studies on Indian Hymenophyllaceae. I. Contributions to our knowledge of *Crepidomanes latealatum* (v.d.B.) Copeland. Proc. Nat. Insti. Sci. Ind. 26 B. 339-351.

———, 1961. Further contributions to our knowledge of *Isoetes sampatkumarani* Rao, L. N. III. J. Indian bot. Soc. **40** : 201-205.

Shende, D. E., 1945. A new species of *Isoetes* from the Bombay Presidency. J. Univ. Bombay **14** : 50-52.

——— and Mahabale, T.S., et al., 1947. Some interesting feature in the morphology of *Isoetes dixitii* Shende from Panchgani, Bombay Presidency. Proc. 34th Indian Sci. Congr. (Delhi).

Singh, T. C. N., 1928. A study of the Mussoorie Ferns (Anatomy). Proc. 15th Indian Sci. Congr. (Calcutta) : 233-234. Also his D.Sc. thesis (1936) Lucknow University.

———, 1928. A note on the presence of a sporangium on the indusium of *Cheilanthes*. Proc. 15th Indian Sci. Congr. (Calcutta) : 234.

———, 1929. A note on the presence of a sporangium on the indusium of *Cheilanthes albomarginata*. J. Indian bot. Soc. **8** : 207.

———, 1930. Note on the anatomy of *Psilotum triquetrum*. Sw. Ann. Bot. Lond. **44** : 1016.

———, 1930. Smut-*Entyloma polysporium* (RK) Farl. on *Selaginella chrysocaulos* in India. New Phytol.

———, 1931. The importance of the indusium in the phylogenetic study of ferns. Proc. 18th Indian Sci. Congr. (Nagpur) : 269.

———, 1931. A study of ferns from the Nainital District. Proc. 18th Indian Sci. Congr. (Nagpur) : 268-269.

———, 1931. The origin of the double indusium in ferns. Proc. 18th Indian Sci. Congr. (Nagpur) : 269.

———, 1931. On the forking of the frond in certain ferns from Nainital Hills. Proc. 18th Indian Sci. Congr. (Nagpur) : 269.

———, 1932. On the ventilating system of certain Indian ferns. Proc. 19th Indian Sci. Congr. (Bangalore) : 309-310,

———, 1932. On the occurrence of *Cuscuta* on ferns. Proc. 19th Indian Sci. Congr. (Bangalore) : 332.

———, 1932. Occurrence of peculiar bodies in the rhizomes and petioles of certain Indian ferns. Proc. 19th Indian Sci. Congr. (Bangalore) : 332-333.

———, 1933. Notes on vegetative budding in certain Indian ferns. Proc. 20th Indian Sci. Congr. (Patna) : 10

Sinha, B. N., 1929. The origin and evolution of Archegonium. J. Indian bot. Soc. 7 : 156-167.

Smith, G. M., 1938. Cryptogamic Botany. Vol. 2, New York.

Spring, A., 1841, 1848. Monographie de la famille des Lycopodiacees. Mem. L' Academic Roy. de Belgique, 1841, XV, Pt 1 ; 1848, XXIV, Pt. II.

Startling, M. H., 1888. On the cultivation of ferns from spores. J. Bombay nat. Hist. Soc. 3: 187.

Stewart, J. L., 1869. Punjab Plants (N.O. Filices). Lahore : 265.

Stewart, R. R., 1938. The Ferns of Mussoorie and Dehra Dun. 150th Anniversary Vol. R. Bot. Gdns. Calcutta. Pt. II : 159-172.

———, 1939. Fern flora of Mussoorie. Proc. 26th Indian Sci. Congr. (Lahore) : 126-127.

———, 1945. Ferns of Kashmir. Bull. Torrey bot. Cl. **72** : 399-426.

———, 1951. The ferns of Pahalgam. J. Indian bot. Soc. **30** : 137-142.

Stewart, R.R., 1951-52. Ferns of Rawalpindi District. Punjab (Pakistan). Forest Records 2 : 88-90.

———, 1957. The ferns and fern-allies of West Pakistan and Kashmir. Biologia 3 : 133-64.

Stokey, A.G., 1948. The gametophyte of *Actiniopteris australis* (L. fil.) Link. J. Indian bot. Soc. **27** : 40-49.

Strachey, R., 1882. Lists of Kumaon plants in E. T. Atkinson's Gazetteer, N. W. Province The Himalayan districts (Lycopodiacees). 652.

Sud, S. R., 1934. A preliminary note on the study of *Azolla pinnata.* J. Indian bot. Soc. **13** : 189-196.

Treub, M., 1888. Gametophyte of *Lycopodium.* Ann. Jard. Bot. Buitenz. **7** : 141-150.

———, 1890. Embryo of *Lycopodium* (Protocorm), Ann. Jard. Buitenzorg. **8** : 1-37.

Trimen, H., 1885. "Systematic catalogue of the Flowering Plants and Ferns of Ceylon." Colombo.

———, 1893-1900. Hand book of the Flora of Ceylon. Vols. *I-V* : London.

Trimen, W., 1882. A Ceylon *Isoetes.* Journ. Botany, Lond. **20** : 353-355.

Varadpande, K. V., 1931. On a species of *Equisetum* from Kolhar river, Nagpur. Proc. 18th Indian Sci. Congr. (Nagpur): 267.

Varma, S. C., 1956. Cytology of *Ophioglossum.* Curr. Sci. **25** : 398-399.

Vasisht, B. R., 1927. The comparative anatomy of *Ophioglossum aitchisoni* d' Almeida and *O. vulgatum* L. Proc. 14th Indian Sci. Congr. (Lahore) ; 215.

——— and Bhagat, R., 1927. The comparative anatomy of *Ophioglossum aitchisoni* d' Almeida and *O. vulgatum* L. J. Indian bot. Soc. **6** : 8-30.

Venkatanathan, T. N., 1931. Observations on the megaspore mother cells of *Isoetes coromandelina.* Proc. 18th Indian Sci. Congr. (Nagpur) : 269.

———, 1932. The microsporangia of *I. coromandelina.* Proc. 19th Indian Sci. Congr. (Bangalore) : 310.

———, 1933. Anatomical studies on *Isoetes coromandelina* L. f. Proc. 20th Indian Sci. Congr. (Patna) : 12.

Venkateshwarlu, V., 1944. On the occurrence of *Psilotum triquetrum* Sw. in the Godavari district. Sci. & Cult. **9** : 165.

Wall, G. W., 1873. A catalogue of the ferns indigenous to Ceylon. London.

Wardlaw, C. W. and Mitra, G.C., 1958. The response of the shoot apex of *Dryopteris aristata* (VILL) Druce and of 'detached' meristems of *Onoclea sensibilis* L. and *Matteuccia struthiopteris* Tod. to physiologically active substances. Bull. Bot. Soc. Beng. (Prof. S. P. Agharkar Commemoration Vol.) 12 : 63-84.

Watt, G., 1889-1915. "A Dictionary of Economic Products of India." Vols. I-VII :

———, 1908. Commercial Products of India.

Wealth of India : (Council of Scientific and Industrial Research Publication). Vols. 1, 2, 3, (1949-1954). *Lycopodium, Marsilia,* etc.

Willis, J. C., 1911. Catalogue of the Flowering Plants and Ferns of Ceylon, Colombo.

——— and Smith, J. M., 1911. Corrections and additions to Trimen's Flora of Ceylon.

# INDEX TO PLANT NAMES AND FAMILIES